S Fiorini and R J Wilson

University of Malta and The Open University

Edge-colourings of graphs

Pitman

LONDON · SAN FRANCISCO · MELBOURNE

PITMAN PUBLISHING LIMITED
39 Parker Street, London WC2B 5PB

FEARON-PITMAN PUBLISHERS INC.
6 Davis Drive, Belmont, California 94002, USA

Associated Companies
Copp Clark Ltd, Toronto
Pitman Publishing Co. SA (Pty) Ltd, Johannesburg
Pitman Publishing New Zealand Ltd, Wellington
Pitman Publishing Pty Ltd, Melbourne

First published 1977

AMS Subject Classifications: (main) 05-XX
(subsidiary) 05Cxx

British Library Cataloguing in Publication Data

Fiorini, S
    Edge colourings of graphs.-(Research notes
    in mathematics; no. 17).
    1. Graph theory
    I. Title II. Wilson, Robin James III. Series
    511'.5   QA166
ISBN 0-273-01129-4

Reproduced and printed by photolithography
in Great Britain at Biddles of Guildford

Preface

In the past hundred years, problems relating to the colouring of the vertices or the regions of graphs and maps have received much attention in the mathematical literature - mainly because of one problem, the four-colour problem. In view of this, it is somewhat surprising that various closely-related problems involving the colouring of the edges of a graph have, until recently, received comparatively little attention.

In this book, we survey the literature of the subject of edge-colourings, and describe some more recent results. Many of these results are rather difficult to locate, and some of the most important papers in the field are available only in Russian. With this book, the subject should become more accessible to those interested in the colouring of graphs.

Special features of this book include: a lengthy introductory section outlining the history of the subject, and summarizing the background material to be assumed in the rest of the book; a set of exercises at the end of each chapter, to test the reader's understanding and to introduce further results; a chapter describing some applications of the material in the Sciences and Social Sciences; an extended bibliography comprising more than 200 items; a very full index; and almost 150 diagrams. The text is divided into four parts:

1. Introduction (3 chapters)
2. The chromatic index (6 chapters)
3. Critical graphs (6 chapters)
4. Further topics (4 chapters).

Finally, we should like to express our thanks to Lowell Beineke, Anthony Hilton, and Ivan Jakobsen, for many useful conversations; to Douglas Woodall for his many comments which resulted in substantial improvements to the text; to the Open University and the University of Malta for their support during the preparation of the manuscript; to Tony Mould for his excellent diagrams; and to Biga Weghofer and the Staff of Pitman Publishing for their help in the publication of this book.

Contents

Part I—Introduction

In these introductory chapters we provide the necessary background and motivation for what follows. We start, in Chapter 1, by giving an historical account of the development of the theory of edge-colourings, from P.G. Tait's 1880 paper to the recent work of V.G. Vizing and others. In Chapter 2, we give the definitions of all of the standard graph-theoretical terms used throughout the remainder of the book. Since several of the properties of edge-colourings included in later chapters are direct analogues or extensions of results relating to the colouring of the vertices or the faces of a graph, we devote Chapter 3 to a summary of these results. It is hoped that by including this introductory material, we have made our book as self-contained as possible.

1 An historical introduction

The origins of chromatic graph theory may be traced back to 1852 when
Augustus De Morgan wrote a letter to his friend, William Rowan Hamilton,
telling him that one of his students had observed that in colouring a map of
England only four colours were needed in order to ensure that neighbouring
countries can be assigned different colours. The 'four-colour conjecture' -
that every map can be coloured with four colours in this way - did not
interest Hamilton, and was almost forgotten until the 1860s when C.S. Peirce
presented a 'proof' in a seminar at Harvard. The conjecture rose to
prominence in 1878 when Arthur Cayley, at a meeting of the London Mathematical
Society, asked whether or not it had been settled. Shortly afterwards,
Cayley published a short note [43] describing the problem, and pointing out
where the difficulties lie. A few months later, in the newly-founded
American Journal of Mathematics, there appeared what purported to be a proof
of the four-colour conjecture; this was A.B. Kempe's well-known fallacious
proof [133], which survived for more than ten years before the error was
discovered. (For further details about the origins of the four-colour
conjecture, the reader is referred to [22].)

   The first papers involving edge colourings appeared in 1880 [175,176]. In
these papers P.G. Tait, Professor of Natural Philosophy in Edinburgh, out-
lined some further 'proofs' of the four-colour conjecture, and deduced that
the edges of every cubic map[†] can be coloured with just three colours in such
a way that the three edges meeting at each vertex are assigned different
colours. He also asserted the converse result without proof, stating
(incorrectly) that one can easily prove by induction that the edges of every
cubic map can be coloured with three colours, thereby giving a simple proof
of the four-colour conjecture. (We shall prove the equivalence of Tait's
edge-colouring result and the four-colour conjecture in Chapter 4.)

---

† All graph-theoretical terms used in this chapter will be explained in
   Chapters 2 and 3.

In 1890 P.J. Heawood published an important paper [100] refuting Kempe's proof, and proving that the countries of every map can be properly coloured with five colours. He followed this eight years later with the first of a series of papers [101-104] in which he took up Tait's ideas, and reformulated them in terms of congruences. The idea of this method is to assign one of the numbers +1 or -1 to each vertex of a cubic map in such a way that the sum of the numbers around each region of the map is congruent to 0 (modulo 3). Heawood proved that the four-colour conjecture is true if and only if every set of congruences obtained in this way has a non-trivial solution.

The next noteworthy result was due to D. König [135]. Following the work of Julius Petersen on the factorization of graphs, König proved that if G is any *bipartite* graph or multigraph and if the maximum valency of G is ρ, then the edges of G can be coloured with exactly ρ colours in such a way that all of the edges meeting at any vertex are coloured differently. We shall prove this result in Chapter 4.

So far, all of the results we have described have involved particular types of graph, such as cubic maps or bipartite graphs. The next result is of a more general nature, and was first proved in an electrical network context by C.E. Shannon in 1949 [166]. It asserts that if G is any graph or multigraph and if the maximum valency of G is ρ, then the edges of G can be coloured using at most $\lfloor \frac{3}{2}\rho \rfloor$ colours. For certain multigraphs this result is best possible, as we shall see in Chapters 4 and 5, but for graphs with no multiple edges it can be improved considerably.

The great breakthrough came in 1964 when V.G. Vizing [191] proved that if G is any graph with maximum valency ρ, then the edges of G can always be coloured with only $\rho + 1$ colours. This magnificent result, which we prove in Chapter 5, generalizes an earlier statement of E.I. Johnson [130] that the edges of every cubic graph can be coloured with only four colours, and shows us that every graph with maximum valency ρ can be classified into one of two classes, according as the number of colours needed is ρ or $\rho+1$. The general problem of deciding which graphs belong to which class is, however, an exceedingly difficult one, and is the problem with which we shall primarily be concerned for much of this book. One of the main methods of attack on this 'classification problem' was introduced by Vizing in two further papers [194,195] and involves the use of so-called 'critical

graphs'. Such graphs have been studied in considerable detail since
Vizing's paper, and we shall describe their properties and their uses in
Parts III and IV of this book.

In 1976 the four-colour problem, the problem which first gave rise to the
study of edge-colourings, was finally settled. Using the method of
'reducible configurations' and a substantial amount of computer time, K.
Appel and W. Haken [5] successfully proved the four-colour conjecture,
thereby proving Tait's result that the edges of every cubic map can indeed be
coloured using only three colours.

2 Basic definitions and examples

In this chapter we shall list the basic graph-theoretic terms we shall be using. Further explanation of these terms can be found in any of the standard texts in graph theory (see, for example, [12,99,200]), although some of the terminology has not yet been completely standardized. Definitions which are not included in this chapter will be introduced as they are needed.

Graphs and Multigraphs

We start by defining a graph G to be a pair (V(G),E(G)), where V(G) is a finite non-empty set of elements called vertices, and E(G) is a finite set of distinct unordered pairs of distinct elements of V(G) called edges (see Figure 2.1); the corresponding object in which the edges may occur several times is called a multigraph (see Figure 2.2). We call V(G) the

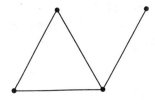
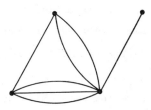

Fig. 2.1   Fig. 2.2

vertex-set of the graph or multigraph G, and E(G) the edge-set of G. Throughout this book the order of G, that is, the number of vertices of G, will be denoted by n; the number of edges will be denoted by m. If e = {v,w} is an edge, then e is said to join the vertices v and w, and these vertices are then said to be adjacent; in this case, we also say that w is a neighbour of v, and that e is incident to v and w. (For convenience, the edge joining v and w will usually be denoted by vw.)

If M is a multigraph, then two or more edges of M joining the same pair of vertices v and w are called multiple edges; the number of such edges is called the edge-multiplicity of vw, denoted by $\mu(vw)$. If M is a multigraph, its underlying graph is the graph $G(M)$ obtained by replacing each set of multiple edges by a single edge; for example, the underlying graph of the multigraph in Figure 2.2 is the graph in Figure 2.1.

For each vertex v of a graph or multigraph G, the number of edges incident to v is called the valency of v, denoted by $\rho(v)$. The maximum valency in a graph G is denoted throughout this book by $\rho(G)$, or simply by ρ when there is no danger of confusion. A vertex of valency 0 is called an isolated vertex. If all the vertices of a graph G have the same valency (ρ, say), then G is called a regular graph, or ρ-valent graph. A 0-valent graph (that is, one with no edges) is often called a null graph, and a 3-valent graph is called a cubic graph. We shall also use the total deficiency of a graph G, which is the sum $\Sigma \{\rho(G) - \rho(v)\}$, where the summation extends over all vertices v of G; if G has n vertices, m edges, and maximum valency ρ, then the total deficiency of G is simply $n\rho - 2m$.

Two edges in a graph G are said to be adjacent if they have a vertex in common. An independent set of edges, or matching, in G is a set of edges of G no two of which are adjacent. The size of the largest independent set of edges in G is called the edge-independence number of G, denoted by α. An independent set of edges which includes every vertex of G is called a 1-factor, or complete matching in G.

Two graphs are said to be isomorphic if there is a one-one correspondence between their vertex-sets which preserves the adjacency of vertices. An automorphism of a graph G is a one-one mapping Φ of $V(G)$ onto itself with the property that $\Phi(v)$ and $\Phi(w)$ are adjacent if and only if v and w are. A subgraph of a graph $G = (V(G), E(G))$ is a graph $H = (V(H), E(H))$ such that $V(H) \subset V(G)$ and $E(H) \subset E(G)$. A subgraph is called a spanning subgraph if $V(H) = V(G)$. If W is any set of vertices of G, then the subgraph induced by W is the subgraph of G obtained by taking the vertices in W and joining those pairs of vertices in W which are joined in G. An induced subgraph of G is a subgraph which is induced by some subset W of $V(G)$. (All of these concepts extend to multigraphs in the obvious way.)

New Graphs from Old

If e is an edge of a given graph G, we use the notation G-e to indicate the graph obtained from G by removing the edge e; more generally, we
write $G - \{e_1, \ldots, e_k\}$ for the graph obtained from G by removing the edges
$e_1, \ldots, e_k$. Similarly, if v is a vertex of G, we use the notation G-v to
indicate the graph obtained from G by removing the vertex v together with
all the edges incident to v; more generally, we write $G - \{v_1, \ldots, v_k\}$ for
the graph obtained from G by removing the vertices $v_1, \ldots, v_k$ and all
edges incident to any of them. We shall also use the notion of <u>contracting a</u>
<u>subgraph to a vertex</u> in which we take a subgraph of G and replace it by a
single vertex adjacent to all those vertices of G which are adjacent to at
least one vertex of the subgraph. We can also talk about <u>inserting a vertex</u>
z <u>into an edge</u> vw of G, where we replace the edge vw by two new edges
vz and zw (see Figure 2.3). If two graphs can be obtained from the same

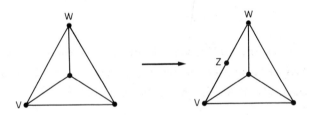

Fig. 2.3

graph by inserting vertices into its edges, the two graphs are called
<u>homeomorphic</u>. The <u>complement</u> $\bar{G}$ of G is the graph with the same vertex-
set as G, but where two vertices are adjacent if and only if they are not
adjacent in G. The <u>line-graph</u> L(G) of G is the graph whose vertices
correspond to the edges of G, and where two vertices are joined if and only
if the corresponding edges of G are adjacent.

If we are given two graphs G and G' whose vertices are labelled, then
their <u>intersection</u> G ∩ G' is the graph with vertex-set V(G) ∩ V(G') and
edge-set E(G) ∩ E(G'). Similarly, their <u>union</u> G ∪ G' is the graph with
vertex-set V(G) ∪ V(G') and edge-set E(G) ∪ E(G'). If V(G) and V(G')
are disjoint, then G ∪ G' is called the <u>disjoint union</u> of G and G'.
But for the purposes of this book, the most useful form of union is the less

7

familiar Hajós-union of G and G' (sometimes called their 'conjunction'); this is obtained by

(i)   choosing a vertex v in G and a vertex v' in G' and identifying them,

(ii) removing any edge vz incident to v and any edge v'z' incident to v', and

(iii) joining the vertices z and z' (see Figure 2.4).

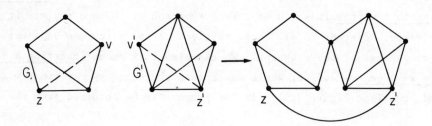

Fig. 2.4

(Note that we can get several different Hajós-unions from G and G' corresponding to the choices involved in steps (i) and (ii); it will always be clear from the context which Hajós-union is intended.)

Paths and Circuits

A sequence of distinct edges of the form $v_0 v_1, v_1 v_2, \ldots, v_{r-1} v_r$ (sometimes written $v_0 \to v_1 \to \ldots \to v_r$) is called a <u>path of length</u> r from v_0 to v_r. If the vertices $v_0, v_1, \ldots, v_r$ are all distinct, then the path is called a <u>chain</u> (or <u>open chain</u>), whereas if the vertices are all distinct except for v_0 and v_r, which coincide, then the path is called a <u>circuit</u> (or <u>closed chain</u>). The length of any shortest open chain from v_0 to v_r is called the <u>distance</u> between v_0 and v_r, and the largest distance between two vertices in a graph G is called the <u>diameter</u> of G. A circuit of length 3 is called a <u>triangle</u>. The length of a shortest circuit in a graph G is called the <u>girth</u> of G, and the length of a longest circuit in G is called the <u>circumference</u> of G. If G has a circuit which includes every vertex in V(G), then such a circuit is called a <u>Hamiltonian circuit</u>, and G is called a <u>Hamiltonian graph</u>. The graph which consists of a single circuit n is called the <u>circuit graph on n vertices</u>, and is denoted by C_n.

Connectivity

A graph G is _connected_ if there is a path joining each pair of vertices (or, equivalently, if G cannot be expressed as the union of two disjoint graphs); a graph which is not connected is called _disconnected_. Clearly, every disconnected graph can be split up into a number of connected subgraphs, and these subgraphs are called _components_. If G is a connected graph, and if the graph G-e is disconnected, for some edge e, then e is called a _bridge_ of G. Similarly, if G is a connected graph, and if the graph G-v is disconnected, for some vertex v, then v is called a _cut-vertex_ of G. We say that a graph G is _k-connected_ if the removal of any set of k-1 vertices does not disconnect the graph; by a famous theorem of Menger (see, for example, [99, p. 47]) this is equivalent to saying that given any pair of non-adjacent vertices v and w in G, there are at least k chains from v to w which are vertex-disjoint (that is, they have no vertices in common except for v and w). The _connectivity_ of G is then defined to be the largest value of k for which G is k-connected. In an analogous way, we can define a graph G to be _k-edge-connected_ if the removal of any set of k-1 edges does not disconnect G; the _edge-connectivity_ of G is then defined to be the largest value of k for which G is k-edge-connected.

Examples of Graphs

A graph in which every two vertices are adjacent is called a _complete graph_; the complete graph with n vertices and $\frac{1}{2}n(n-1)$ edges is denoted by K_n. A _bipartite graph_ is one whose vertex-set can be partitioned into two sets in such a way that each edge joins a vertex of the first set to a vertex of the second set. A _complete bipartite graph_ is a bipartite graph in which every vertex in the first set is adjacent to every vertex in the second set; if the two sets contain r and s vertices respectively, then the complete bipartite graph is denoted by $K_{r,s}$. Any complete bipartite graph of the form $K_{1,s}$ is called a _star graph_. A _complete r-partite graph_ is obtained by partitioning the vertex-set into r sets, and joining two vertices if and only if they lie in different sets. The _Petersen graph_ is the graph shown in Figure 2.5, and the graphs corresponding to the vertices and edges of the five regular solids are called the _platonic graphs_ (see Figure 2.6 for the graphs of the cube and the octahedron).

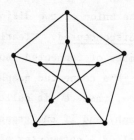 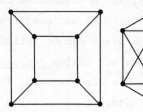 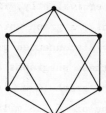

Fig. 2.5   Fig. 2.6

A planar graph is a graph which can be embedded in the plane in such a way that no two edges intersect geometrically except at a vertex to which they are both incident. If a connected graph G is embedded in the plane in this way, it is called a plane graph; in this case, the points of the plane not on G are partitioned into open regions called faces, and the number f of such faces is given by Euler's polyhedral formula

$$n - m + f = 2,$$

where n and m denote (as usual) the number of vertices and edges of G. An outerplanar graph is a planar graph which can be embedded in the plane in such a way that every vertex lies on the unbounded face (see Figure 2.7). A connected planar graph which contains no bridges is often called a map.

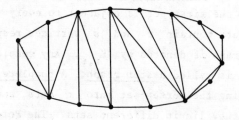

Fig. 2.7

<u>And finally...</u>

If S is a finite set, we denote the number of elements in S by $|S|$; the empty set will be denoted by $\emptyset$. We shall use $[x]$ for the largest integer not greater than x, and $\{x\}$ for the smallest integer not smaller than x (so that, for example, $[\pi] = 3$, $\{\pi\} = 4$). The end (or absence) of a proof will be denoted by the symbol $\square$.

3 Vertex-colourings of graphs

In this chapter we shall describe some of the main results in classical chromatic graph theory, concentrating in particular on those results which are of relevance to the rest of this book. Some of these will turn out to have natural analogues in the theory of edge-colourings, whereas others are included because they will be needed in later proofs. We shall sometimes be content in this chapter simply to state the results we need; their proofs may easily be found in any of the standard textbooks on the subject.

The Chromatic Number

If G is a graph, we define its <u>chromatic number</u> $\chi(G)$ to be the minimum number of colours needed to colour the vertices of G in such a way that no two adjacent vertices are assigned the same colour. If $\chi(G) = k$, we say that G is <u>k-chromatic</u>, and if $\chi(G) \le k$, we say that G is <u>k-colourable</u> (or <u>k-colourable(v)</u>, if there is any possibility of confusion with face- or edge-colourings). For example, the complete graph K_n is n-chromatic, every null graph is 1-chromatic, and the circuit graph C_n is 2-chromatic or 3-chromatic according as n is even or odd. Note that if G is a bipartite graph, then G is 2-colourable.

If we know the largest valency ρ in a graph G, then we can easily give an inductive argument to prove that G is necessarily $(\rho+1)$-colourable. Using a 'Kempe-chain argument' (to be described below), this result can be strengthened to give the following result, known as Brooks' theorem [34]; its proof may be found in [12, p.202]:

<u>Theorem 3.1</u> (Brooks). Let G be a connected graph which is not a complete graph or a circuit of odd length, and let ρ be the largest valency in G. Then G is ρ-colourable. □

Vertex-Critical Graphs

In studying the chromatic number of arbitrary graphs, it is often useful
to restrict oneself to graphs which are critical in some sense. For example,
in studying k-chromatic graphs in general, one can often restrict one's
attention to graphs which are k-chromatic, but 'only just', in the sense that
although G needs k colours, any subgraph of G can be coloured with
fewer than k colours. By restricting ourselves to critical graphs in this
way we lose nothing, and often gain a lot, since critical graphs generally
turn out to have more structure than arbitrary graphs.

More precisely, we define a graph G to be vertex-critical(v) if
$\chi(G-v) < \chi(G)$ for each vertex v of G. For example the graph G in
Figure 3.1 is vertex- critical(v), since $\chi(G) = 4$, whereas $\chi(G-v) \leq 3$ for
each vertex v.

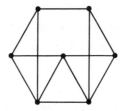

Fig. 3.1

The following properties of vertex-critical graphs were proved by Dirac
[60]; we shall need to refer to them when we study (edge-) critical graphs
in Part III:

Theorem 3.2. Let G be a vertex-critical(v) graph with $\chi(G) = k$. Then
(i)   G is a connected graph;
(ii)   the valency of each vertex of G is at least k-1;
(iii)   G cannot be expressed in the form $G_1 \cup G_2$, where G_1 and G_2 are
        graphs which intersect in a complete graph; in particular, G
        contains no cut-vertices.

Proof
(i)   If G is a disconnected graph, let C be any component of G with
chromatic number k, and let v be any vertex of G which is not in C.
Then $\chi(G-v) = k$, contradicting the fact that G is vertex-critical.

(ii) Since G is vertex-critical, we have $\chi(G-v) \leq k-1$ for each
vertex v. If the valency of v is less than k-1, then the neighbours of
v will be coloured with at most k-2 colours. It follows that any (k-1)-
colouring of G-v can be extended to a (k-1)-colouring of G, contradicting
the fact that $\chi(G) = k$.

(iii) If $G = G_1 \cup G_2$ and $G_1 \cap G_2 = K_r$, then $\chi(G_1) \leq k-1$ and $\chi(G_2) \leq k-1$,
since G_1 and G_2 are subgraphs of G, and G is critical. But by
relabelling the colours, we can ensure that the vertices in $G_1 \cap G_2$ are
coloured the same way in both graphs. These two colourings can then be
combined to give a (k-1)-colouring for G, contradicting the fact that
$\chi(G) = k$. By considering the case r = 1, it follows immediately that G
can contain no cut-vertices. □

The Four-Colour Theorem

Although the chromatic number of an arbitrary graph cannot be estimated
at all accurately, the opposite situation holds for a planar graph. Before
deriving any results of this kind, we shall first obtain a version of
Euler's polyhedral formula which will be useful both here and in Chapter 16.

Theorem 3.3. Let G be a connected planar graph with n ($\geq$ 3) vertices.
If G has exactly n_k vertices of valency k, for each $k \geq 1$, then
$$5n_1 + 4n_2 + 3n_3 + 2n_4 + n_5 - n_7 - 2n_8 - \ldots \geq 12,$$
with equality if and only if every face of G is a triangle.

Proof. If G has n vertices and m edges, then clearly
$$n = n_1 + n_2 + n_3 + \ldots$$
and
$$2m = n_1 + 2n_2 + 3n_3 + \ldots .$$
Also, since every face of G is bounded by at least three edges, we have
$3f \leq 2m$, with equality if and only if every face of G is a triangle. The
required results now follow by substituting these expressions into Euler's
polyhedral formula $n - m + f = 2$. □

It follows from Theorem 3.3 that $5n_1 + 4n_2 + \ldots + n_5$ must necessarily
be positive, and we can therefore deduce the following corollary:

Corollary 3.4. Every planar graph has a vertex whose valency is at most five. □

We can now prove the following result, usually known as the 'five-colour theorem'.

Theorem 3.5. Every planar graph is 5-colourable.

Proof. We use induction on the number of vertices of the graph, and assume the theorem to be true for all planar graphs with at most n vertices.

Let G be a planar graph with n+1 vertices. By Corollary 3.4, G contains a vertex v whose valency is at most five. The graph G-v is a planar graph with n vertices, and so can be coloured with five colours, by the induction hypothesis. Our aim is to show how this colouring of the vertices of G-v can be modified to give a colouring of the vertices of G. We may assume that v has exactly five neighbours, and that they are differently coloured, since otherwise there would be at most four colours adjacent to v, leaving a spare colour which could be used to colour v; this would complete the colouring of the vertices of G. So the situation is now as in Figure 3.2, with the vertices $v_1, \ldots, v_5$ coloured $\alpha, \beta, \gamma, \delta, \varepsilon$, respectively.

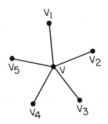

Fig. 3.2

If λ and μ are any two colours, we define $H(\lambda, \mu)$ to be the two-coloured subgraph of G induced by all those vertices coloured λ or μ. We shall first consider $H(\alpha, \gamma)$; there are two possibilities:

(i) If v_1 and v_3 lie in different components of $H(\alpha, \gamma)$ (see Figure 3.3), then we can interchange the colours α and γ of all the vertices in the component of $H(\alpha, \gamma)$ containing v_1. The result of this re-colouring is that v_1 and v_3 both have colour γ, enabling v to be coloured α. This completes the proof in this case.

15

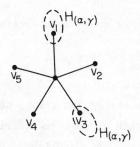

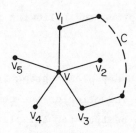

Fig. 3.3   Fig. 3.4

(ii) If v_1 and v_3 lie in the same component of $H(\alpha,\gamma)$ (see Figure 3.4), then there is a circuit C of the form $v \to v_1 \to \ldots \to v_3 \to v$, the part between v_1 and v_3 lying entirely in $H(\alpha,\gamma)$. Since v_2 lies inside C and v_4 lies outside C, there cannot be a two-coloured chain from v_2 to v_4 lying entirely in $H(\beta,\delta)$. We can therefore interchange the colours of all the vertices in the component of $H(\beta,\delta)$ containing v_2. The vertices v_2 and v_4 are both now coloured δ, enabling v to be coloured β. This completes the proof. ☐

The argument used in the proof of Theorem 3.5 (namely that of looking at a two-coloured subgraph $H(\alpha,\gamma)$ and interchanging the colours) is often called a Kempe-chain argument, since it was initiated by A.B. Kempe in 1879 in his abortive attack on the four-colour conjecture [133]. We shall be using such Kempe-chain arguments when we come to edge-colourings of graphs, the only difference being that in the edge-colouring case each two-coloured subgraph $H(\alpha,\beta)$ has a particularly simple form - namely, a union of open chains and circuits.

Kempe's original proof of the four-colour conjecture was incorrect, since its author attempted to perform two colour-interchanges simultaneously. That this does not work was pointed out by Heawood in 1890 [100], and it was not until 1976 that the problem was finally settled. K. Appel and W. Haken used the theory of reducible configurations, developed by G.D. Birkhoff in 1913 [24], together with the theory of unavoidable sets (see Heesch [105]), to obtain an unavoidable set containing nearly two thousand reducible configurations. For further details, the reader is referred to [5], but for convenience we shall state the result here:

Theorem 3.6 (The Four-Colour Theorem). Every graph is 4-colourable. □

Duality

   If G is any connected planar graph, we define a dual G* of G in the following way: (i) take a plane embedding of G, and place one point in each face of G - these points are the vertices of G*; (ii) for each edge e of G, draw a line crossing e and joining the two vertices of G* which lie in the faces separated by e - these lines are the edges of G* (see Figure 3.5). It is clear that this procedure can be carried out for any

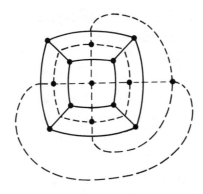

Fig. 3.5

planar graph, and we can easily deduce the following elementary consequences (see, for example, [200, pp. 73-74]):

Theorem 3.7. Let G be a plane connected graph, and let G* be a dual of G. Then
(i)   (G*)* is isomorphic to G;
(ii) if G has n vertices, m edges and f faces, then G* has f vertices, m edges and n faces. □

   Interest in colouring graphs first arose out of problems involving the colouring of the countries (i.e. faces) of a map. With this in mind, we define a map G to be k-colourable(f) if its faces can be coloured with k colours in such a way that no two adjacent faces (that is, faces with a

boundary edge in common) have the same colour. The following theorem can
then be deduced immediately from the definition of G* (see, for example,
[200, p.90]):

Theorem 3.8. Let G be a planar graph, and let G* be a dual of G; then
G is k-colourable(v) if and only if G* is k-colourable(f). □

    Using this result, we can restate the four-colour theorem in terms of
face-colourings:

Corollary 3.9. Every map is 4-colourable(f). □

    We conclude this chapter with a very simple result involving face-
colourings. It will be used in the proof of Tait's Theorem (Theorem 4.4) and
we leave its proof to the reader:

Theorem 3.10. If G is a regular 2-valent graph (that is, a union of disjoint circuits), then G is 2-colourable(f). □

Exercises

3a Find the chromatic number of (i) the Petersen graph, (ii) each of the
    platonic graphs. Are any of these graphs vertex-critical(v)?

3b A graph G is edge-critical(v) if $\chi(G-e) < \chi(G)$ for each edge e.
    (i)   Prove that if G is edge-critical, then G is also vertex-
           critical.
    (ii)   By considering the graph of Figure 3.1, show that the converse of
           part (i) is false.
    (iii) Prove that if G_1 and G_2 are both edge-critical and k-chromatic,
           then so is any Hajós-union of G_1 and G_2.

3c (i)   Prove that a map G is 2-colourable(f) if and only if every vertex
           of G has even valency.
    (ii)   Prove that a cubic map G is 3-colourable(f) if and only if every
           face is bounded by an even number of edges.

3d Let χ be the chromatic number of a graph G with n vertices, and let
    $\bar{\chi}$ be the chromatic number of its complement $\bar{G}$. Prove that

18

$$2\sqrt{n} \;\leq\; \chi + \bar{\chi} \;\leq\; n + 1,$$

and

$$n \;\leq\; \chi\bar{\chi} \;\leq\; \tfrac{1}{4}(n+1)^2.$$

Show that these results are best possible.   (Nordhaus and Gaddum [156].)

3e   The <u>clique number</u> $\omega(G)$ of a graph G is the order of the largest
     complete subgraph in G (so that if K_r is a subgraph of G, but
     K_{r+1} is not, then $\omega(G) = r$). Prove that if G is a planar graph,
     then $\chi(G) \leq \omega(G) + 1$.

3f   A graph is <u>uniquely colourable(v)</u> if its vertex-set can be partitioned
     in only one way into vertices of the same colour. Prove that if G is
     a uniquely colourable graph with $\chi(G) = k$, then
     (i)   the valency of each vertex of G is at least k-1;
     (ii)   for any two colours α and β, the subgraph $H(\alpha,\beta)$ induced by
            the vertices coloured α or β is connected.

3g   (i)   Let G be a graph with vertices $v_1, v_2, \ldots, v_n$ and let A be the
            adjacency matrix of G (that is, the matrix $A = (a_{ij})$, where
            $a_{ij} = 1$ if v_i and v_j are adjacent, and 0 otherwise). If r
            is the maximum eigenvalue of A, prove that $\chi(G) \leq 1 + r$, with
            equality if and only if G is a complete graph or a circuit of odd
            length.
     (ii)   Compare the upper bound of part (i) with that given by Brooks'
            Theorem when G is the star graph with n vertices.
                                              (Szekeres and Wilf [174].)

3h   Prove that if G is an infinite graph, every finite subgraph of which
     is k-colourable, then G is k-colourable.   (de Bruijn and Erdös [48].)

Part II—The chromatic index

In this part we begin our study of the edge-colourings of graphs. We start, in Chapter 4, by defining the chromatic index of a graph or multigraph, and deriving the value of this parameter for certain important types of graph. In the following chapter we prove the fundamental theorem of Vizing, mentioned in Chapter 1, and use the multigraph extension of this theorem to obtain a proof of Shannon's electrical network result. Using Vizing's Theorem we can classify graphs into two classes, depending on the value of their chromatic index, and this classification problem is discussed in Chapter 6. This discussion spills over into Chapter 7, where we describe various generalizations of the Petersen graph.

The last two chapters in Part II are of a somewhat different nature, and the results contained therein will not be used in later chapters. In Chapter 8 we discuss regular graphs, and describe some of the connections which exist between the theory of edge-colourings and group theory or matrix theory. Finally, in Chapter 9, we show how results on edge-colourings are of use in electrical network theory, the design of experiments, and statistics; we also show how certain results of edge-colourings can be proved using ideas of tensor calculus.

4 The chromatic index

In this chapter we define the chromatic index of a graph or multigraph, and illustrate this definition by determining the chromatic index of certain types of graph. In particular, we determine the chromatic index of the complete graphs and the complete bipartite graphs, and present two proofs of König's result on the chromatic index of a general bipartite graph. The chapter concludes with a proof of Tait's theorem on cubic maps, referred to in Chapter 1.

The Chromatic Index

If G is a graph or multigraph, we define its <u>chromatic index</u> $\chi'(G)$ to be the minimum number of colours needed to colour the edges of G in such a way that no two adjacent edges are assigned the same colour. (The chromatic index is sometimes referred to in the literature as the edge-chromatic, or line-chromatic, number.) For example, if C_n is the circuit graph with n (≥ 3) vertices, then $\chi'(C_n) = 2$ if n is even, and $\chi'(C_n) = 3$ if n is odd. If k is any integer satisfying $\chi'(G) \leq k$, then G is said to be <u>k-edge-colourable</u> (or simply <u>k-colourable</u>, when there is no possibility of confusion). So, for example, every circuit graph C_n is 3-colourable. If the edges of G have been coloured with k colours, then any maximal set of edges of the same colour is called a <u>colour class</u>.

It is clear from the definition that if G contains a vertex of valency t, then $\chi'(G) \geq t$, and it follows that the maximum valency of G is necessarily a lower bound for the chromatic index. It is also clear from the definition that if G is a graph with at least one edge, then the chromatic *index* of G is equal to the chromatic *number* of $L(G)$, the line-graph of G. For example, the chromatic index of the complete graph K_4 is equal to the chromatic number of its line-graph, the graph of the octahedron (see Figure 2.6). We shall occasionally make use of this relationship between $\chi'(G)$ and $\chi(L(G))$ to obtain results on the chromatic index from corresponding known results on the chromatic number, although it usually turns out to be more useful to deal with the chromatic index directly.

Complete Graphs and Complete Bipartite Graphs

The chromatic index of the complete graph K_n has been determined by several authors using a variety of methods (see, for example, Vizing [194], Behzad, Chartrand and Cooper [13], and Berge [17, p.249]). Our proof will be mainly constructive.

Theorem 4.1. The chromatic index of K_n ($n \geq 2$) is given by $\chi'(K_n) = n$, if n is odd, and $\chi'(K_n) = n-1$, if n is even.

Proof. We note first that $\chi'(K_n) \geq n-1$, since every vertex of K_n has valency $n-1$.

If n is odd, then the maximum possible number of edges which can be coloured with the same colour is precisely $\frac{1}{2}(n-1)$, since otherwise two of these edges would be adjacent. It follows that K_n has at most $\frac{1}{2}(n-1) \chi'(K_n)$ edges, and hence that $\chi'(K_n) \geq n$. To prove equality, we can either quote Vizing's theorem (see Chapter 5), or we can explicitly construct an n-colouring of the edges. To effect such a colouring, we place the vertices of K_n in the form of a regular n-gon, and colour the edges around the boundary using a different colour for each edge. The remaining edges can then be coloured by assigning to each one the same colour as that used for the boundary edge parallel to it. Figure 4.1 illustrates the case n=5.

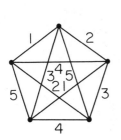

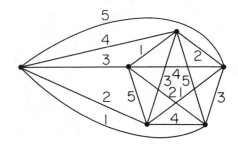

Fig. 4.1   Fig. 4.2

If n is even, we can prove that $\chi'(K_n) = n-1$ by explicitly constructing an (n-1)-colouring of the edges of K_n. If $n = 2$, this is trivial. If $n > 2$, we choose any vertex v, and colour the edges of $K_n - v$

(a complete graph on n-1 vertices) in the manner described above. At each
vertex there will be exactly one colour missing, and these missing colours
will all be different. The edges of K_n incident to v can then be
coloured using these missing colours (see Figure 4.2 for the case n=6). ☐

We can interpret the result of Theorem 4.1 in the following way. Suppose
that n teams take part in a competition in which each team is required to
play exactly one match against each of the other teams. If we assume that
any matches which involve different pairs of teams may be played
simultaneously, it follows immediately that the minimum number of contests
necessary for all the games to be played is n, if n is odd, and n-1, if
n is even.

The chromatic index of the complete bipartite graph $K_{r,s}$ is also easy
to determine. We shall again use an explicit construction.

<u>Theorem 4.2.</u> The chromatic index of $K_{r,s}$ is given by $\chi'(K_{r,s}) = \max\{r,s\}$.
<u>Proof.</u> We may assume, without loss of generality, that $r \geq s$; it follows
immediately that $\chi'(K_{r,s}) \geq r$. To prove equality, we suppose that $K_{r,s}$ is
drawn with the s vertices in a horizontal line below the r vertices (see
Figure 4.3 which represents $K_{4,3}$). The required r-colouring is then

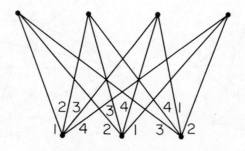

Fig. 4.3

effected by successively colouring the edges incident to these s vertices
in a clockwise direction, using the colours
$$\{1,2,\ldots,r\};\{2,3,\ldots,r,1\}; \ \ldots \ ;\{s,\ldots,r,1,\ldots,s-1\}. \quad ☐$$

König's Theorem for Bipartite Graphs

In fact, Theorem 4.2 is a special case of an important result for general
bipartite graphs, which states that if G is any bipartite graph with
maximum valency ρ, then χ'(G) = ρ. This was first proved by Dénes König
[135] in 1916, in connection with the factorization of graphs. König's
proof uses induction on the number of edges of the graph, and involves a
Kempe-chain argument; this type of argument was discussed in Chapter 3, and
involves looking at a two-coloured subgraph H(α,β) and interchanging these
colours. (A more complicated example of a Kempe-chain argument will be
presented in the next chapter.) We shall also present an alternative proof
which relies on the decomposition of a bipartite graph into matchings; this
type of argument is discussed at some length in Berge's book [17], to which
the reader is referred for further details.

Theorem 4.3 (König's Theorem). If G is a bipartite graph or multigraph
with maximum valency ρ, then χ'(G) = ρ.
First Proof. We use induction on the number of edges of G; it is clearly
sufficient to prove that if all but one of the edges of G have been
coloured using at most ρ colours, then there is a ρ-colouring of all of
the edges of G.

So suppose that each edge of G has been coloured with one of the ρ
given colours, with the single exception of the edge e = vw. Then there
must be at least one colour not assigned to the edges incident to v, and at
least one colour not assigned to those incident to w. If there is some
colour missing at both v and w, then this colour can be used to colour
the edge e and the proof is complete. If this is not the case, let α be
any colour missing at v, let β(≠ α) be any colour missing at w, and let
H(α,β) be the connected subgraph of G consisting of the vertex w and
all those vertices and edges of G which can be reached from w by a path
consisting entirely of edges coloured α or β. Since G is a bipartite
graph, the subgraph H(α,β) cannot contain the vertex v, and so we can
interchange the colours α and β in this subgraph without affecting v or
the rest of the colouring. The edge e can then be coloured α, thereby
completing the colouring of the edges of G.   □

Second Proof. We use induction on the maximum valency of G, and assume the
well-known result (see, for example, Berge [17, p.135]) that every bipartite

graph has a matching which contains all vertices of maximum valency. If M
is a matching which contains every vertex of valency ρ, then the graph G-M
is a bipartite graph with maximum valency $\rho-1$. It follows from our
induction hypothesis that the edges of G-M can be coloured with $\rho-1$
colours. The required ρ-colouring of the edges of G can now be effected
by colouring the edges of M with the ρth colour. ☐

Tait's Theorem

We conclude this chapter with a discussion of those graphs which
initiated the whole subject - cubic maps (see Chapter 1). If G is a
cubic map, then its line-graph L(G) is a regular 4-valent graph which is
not isomorphic to K_5. It follows from Brooks' Theorem (Theorem 3.1) that
$\chi(L(G)) \leq 4$, and hence that $\chi'(G) \leq 4$, a result first proved by E.I.
Johnson [130] in 1963. The fact that $\chi'(G) = 3$ for all cubic maps G
follows as a direct consequence of Tait's classic result of 1880 [175],
relating the chromatic index of cubic maps to the four-colour theorem; it is
because of this theorem that a 3-colouring of the edges of a cubic map is
sometimes referred to as a Tait colouring.

Theorem 4.4 (Tait). The four-colour theorem is equivalent to the statement
that every cubic map has chromatic index 3.
Proof. Suppose first that G is a cubic map whose faces are coloured with
the four colours A, B, C and D. We can obtain a 3-colouring of the edges of
G by colouring with colour 1 those edges which separate faces coloured
A and B, or C and D, by colouring with colour 2 those edges which
separate faces coloured A and C, or B and D, and by colouring with
colour 3 those edges which separate faces coloured A and D, or B and
C (see Figure 4.4). It is easy to see that this colouring assigns the

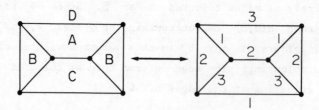

Fig. 4.4

colours 1,2 and 3 to the edges of G in such a way that the edges meeting at each vertex are differently coloured.

To prove the converse result, it is sufficient to prove that if G is any cubic map whose edges have been coloured with the colours 1,2 and 3, then the faces of G can be coloured with four colours. But the subgraph of G determined by those edges coloured 1 or 2 is a regular 2-valent graph whose faces can be coloured with two colours α and β, by Theorem 3.10. In a similar way, the faces of the subgraph determined by those edges coloured 1 or 3 can be coloured with the colours γ and δ. It follows that we can assign to each face of G two coordinates (x,y), where x is either α or β, and y is either γ or δ. Since the coordinates assigned to two adjacent faces of G must differ in at least one place, it follows that these coordinates $A = (\alpha,\gamma)$, $B = (\beta,\delta)$, $C = (\beta,\gamma)$, $D = (\alpha,\delta)$, give the required 4-colouring of the faces of G. □

Corollary 4.5. Every cubic map has chromatic index 3. □

Exercises.

4a Find the chromatic index of each of the platonic graphs.

4b (i)   Show that if P is the Petersen graph, then $\chi'(P) = 4$; show also
          that if v is any vertex of P, then $\chi'(P-v) = 4$.

    (ii) Obtain corresponding results for the graph shown in Figure 4.5.

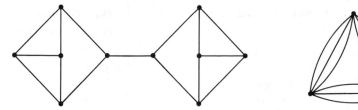

Fig. 4.5   Fig. 4.6

4c (i)   Show that if M is either of the multigraphs in Figure 4.6, then
          $\chi'(M) = [\frac{3}{2}\rho]$.

    (ii) Show, more generally, that if M(k) is the multigraph with three
          vertices mutually joined by $[\frac{1}{2}k]$, $[\frac{1}{2}k]$ and $[\frac{1}{2}(k+1)]$ edges,

27

respectively, then $\chi'(M) = [\frac{3}{2}\rho]$. (These graphs are called the Shannon multigraphs, and show that Shannon's upper bound mentioned in Chapter 1 can be attained for any value of ρ.)

4d Let G be a cubic Hamiltonian graph. Prove that $\chi'(G) = 3$.

4e Let G be a graph (other than an odd circuit) in which every circuit has odd length. Show that if the maximum valency of G is ρ, then $\chi'(G) = \rho$.

4f Let G be a graph with maximum valency ρ. Use Brooks' Theorem (Theorem 3.1) to prove that $\chi'(G) \leq 2\rho-2$.

4g Let G be a cubic map. Use Corollary 4.5 to prove that each vertex of G can be assigned one of the numbers +1 and -1 in such a way that the sum of the numbers around each face is a multiple of 3.

(Heawood [101].)

4h (i)   Let K(k,r) be the complete r-partite graph, each of whose parts has exactly k vertices. (The graph K(2,3) is shown in Figure 4.7.) Prove that the chromatic index of K(k,r) is equal to kr-k+1, if both k and r are odd, and kr-k, otherwise.

(Laskar and Hare [141].)

(ii) Let C(k,r) be the 'generalized circuit graph' obtained by arranging r copies of the null graph with k vertices into a cycle, and joining two vertices if and only if they lie in neighbouring members of the cycle. (The graph C(2,3) is shown in Figure 4.7.) Prove that the chromatic index of C(k,r) is equal to 2k+1, if both k and r are odd, and 2k, otherwise.

(Parker [158].)

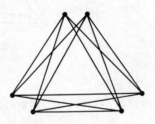

Fig. 4.7

4i Let χ' be the chromatic index of a graph G with n vertices, and
let $\bar{\chi}'$ be the chromatic index of its complement $\bar{G}$. Prove that if n
is even, then

$$n-1 \leq \chi' + \bar{\chi}' \leq 2n-2$$
$$\text{and} \quad 0 \leq \chi'\bar{\chi}' \leq (n-1)^2,$$

and that if n is odd, then

$$n \leq \chi' + \bar{\chi}' \leq 2n-3$$
$$\text{and} \quad 0 \leq \chi'\bar{\chi}' \leq (n-1)(n-2).$$

Show that all of these results are best possible. (Compare these results
with those of Exercise 3d.)   (Vizing [194], Alavi and Behzad [1].)

4j Let G be a graph, and let $\tilde{r}$ be the largest eigenvalue of the
adjacency matrix of its line-graph $L(G)$. Prove that $\chi'(G) \leq 1+\tilde{r}$, with
equality if and only if G is a star graph or a circuit of odd length.
(Compare this result with that of Exercise 3g.)

4k (i)   Prove that if G is an infinite graph, every finite subgraph of
which is k-edge-colourable, then G is k-edge-colourable.
(Compare this result with that of Exercise 3h.)   (Neumann [155].)

(ii) Deduce from part (i) that if G is an infinite bipartite graph
with maximum valency ρ (finite), then G is ρ-edge-colourable.

5 The theorems of Vizing and Shannon

In our historical introduction in Chapter 1 we remarked that C.E. Shannon
[166] had obtained an upper bound of $[\frac{3}{2}\rho]$ for the chromatic index of a
multigraph with maximum valency ρ; this bound can actually be achieved for
any value of ρ, as we saw in Exercise 4c. If, now, we restrict ourselves
to graphs, rather than multigraphs, we can obtain far more restrictive upper
bounds for the chromatic index, and our main aim in this chapter is to prove
the most important of these - namely, Vizing's Theorem, which states that
if G is a graph with maximum valency ρ, then the chromatic index of G
cannot exceed $\rho+1$. We then show how Vizing's Theorem can be extended to
multigraphs, and we conclude by using the multigraph version to prove
Shannon's Theorem.

Vizing's Theorem

The first proof of Vizing's Theorem appeared in 1964 [191,194], and the
result was later proved independently by Gupta [94]. Since then several more
proofs have appeared, and the reader is referred to [12,17,27,62,80,157] for
further details of these proofs. The one we give here depends on a Kempe-
chain argument, and to carry it out we shall need the following notation:
if G is a given graph whose edges are coloured $\lambda,\mu,\dots$, then $H(\lambda,\mu)$
denotes the two-coloured subgraph of G induced by those edges which are
coloured λ or μ; note that, unlike similar proofs involving vertex-
colourings (see Chapter 3), each component of $H(\lambda,\mu)$ must be either an
open chain or a circuit of even length.

Theorem 5.1 (Vizing's Theorem). If G is a graph with maximum valency ρ,
then either $\chi'(G) = \rho$ or $\chi'(G) = \rho+1$.
Proof. It is clear that $\chi'(G) \geq \rho$. To prove that $\chi'(G) \leq \rho+1$, we use
induction on the number of edges of G; more precisely, we shall prove that
if all but one of the edges of G have been coloured with at most $\rho+1$
colours, then there is a $(\rho+1)$-colouring of the edges of G.
So suppose that each edge of G has been coloured with one of the $\rho+1$

given colours, with the single exception of the edge $e_1 = vw_1$. Then there must be at least one colour missing at the vertex v and at least one colour missing at the vertex w_1. If there is some colour missing at both v and w_1, then this colour can be used to colour the edge e_1, and the proof is complete. If this is not the case, let α be any colour missing at v, and let $\beta_1 (\neq \alpha)$ be any colour missing at w_1.

The proof now proceeds in three steps, as follows:

Step 1: Let $e_2 = vw_2$ be the edge incident with v which has been assigned the colour β_1; such an edge must exist, since otherwise the colour β_1 would be missing from both v and w_1, contrary to our assumption. We now un-colour the edge e_2, and assign the colour β_1 to the edge e_1 (see Figure 5.1). We may assume that the vertices v, w_1 and w_2 all belong to

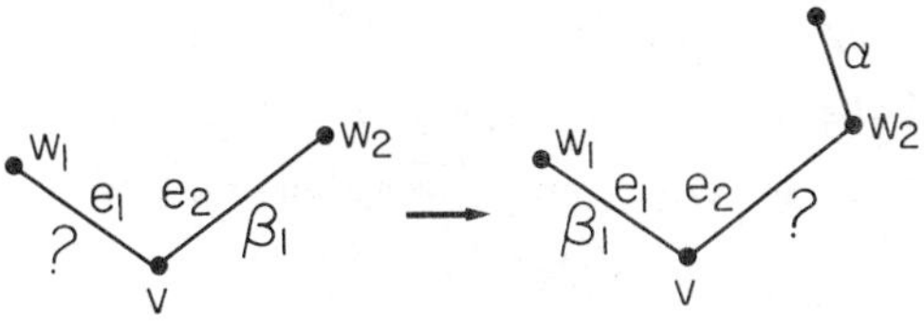

Fig. 5.1 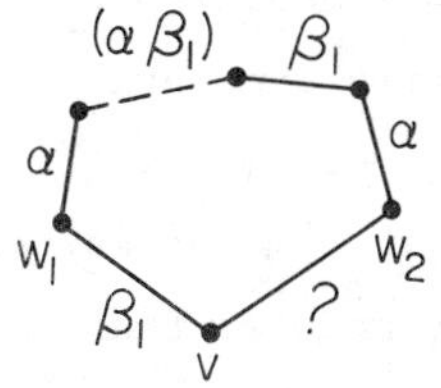   Fig. 5.2

the same component of $H(\alpha, \beta_1)$, since otherwise we could interchange the colours of the edges in the component containing w_2 without altering the colour of e_1; this would mean that the edge e_2 could then be coloured α, thereby completing the colouring of the edges of G. So the situation is now as depicted in Figure 5.2.

Step 2: Now let $\beta_2 (\neq \beta_1)$ be a colour missing at w_2. We may assume that β_2 occurs at v, since otherwise we can complete the proof by using β_2 to colour the edge e_2. So let $e_3 = vw_3$ be the edge incident to v coloured β_2. Then we can un-colour the edge e_3 and assign the colour β_2 to the edge e_2. By the same argument as we used in Step 1, we may assume that the vertices v, w_2 and w_3 all belong to the same component of $H(\alpha, \beta_2)$. The situation is now as depicted in Figure 5.3.

Step 3: If we repeat the above procedure, we shall eventually reach a vertex w_k adjacent to v such that the edge vw_k is un-coloured, and some colour $\beta_i (i < k-1)$ is missing from the vertex w_k. As before, we may assume that

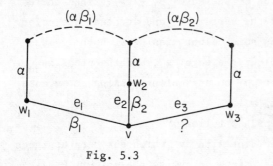

Fig. 5.3

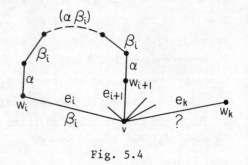

Fig. 5.4

the vertices v, w_i and w_{i+1} all belong to the same component Γ of $H(\alpha, \beta_i)$. Since α does not appear at v, nor β_i at w_{i+1}, Γ must be a chain from v to w_{i+1} passing through w_i, and consisting entirely of edges coloured alternately β_i and α (see Figure 5.4). Clearly this chain does not contain w_k, since β_i does not appear at w_k. It follows that if $\tilde{\Gamma}$ is the component of $H(\alpha, \beta_i)$ containing the vertex w_k, then Γ and $\tilde{\Gamma}$ must be disjoint. It is therefore possible to interchange the colours of the edges in $\tilde{\Gamma}$, thereby enabling us to colour the edge vw_k with colour α. This completes the proof. $\square$

Vizing also obtained a corresponding bound for multigraphs which is some-times (but not always) better than that given by Shannon. It involves the maximum multiplicity μ of the multigraph M - that is, the maximum of all the various edge-multiplicities in M; for example, the maximum multiplicities of the two 'Shannon multigraphs' in Figure 4.6 are 3 and 4, respectively. The multigraph version of Vizing's Theorem may now be stated as follows:

Theorem 5.2. If M is a multigraph with maximum valency ρ and maximum multiplicity μ, then

$$\rho \le \chi'(M) \le \rho + \mu. \quad \square$$

The upper bound given by Theorem 5.2 was also obtained independently by Gupta [94], and is clearly an improvement on Shannon's result whenever M is a multigraph for which $\mu < \frac{1}{2}\rho$. A further slight improvement has been given by Hilton [107] in the case when *every* two adjacent vertices are joined by exactly μ edges; such a multigraph is called a μ-uniform multigraph:

Theorem 5.3. If M is a μ-uniform multigraph with maximum valency ρ, and if $G(M)$, the underlying graph of M, has at most $\mu-1$ edges, then

$$\chi'(M) < \rho + \mu, \text{ if } G(M) \neq K_{2k+1} \text{ for any } k \geq 1,$$
$$\text{and } \chi'(M) = \rho + \mu, \text{ if } G(M) = K_{2k+1} \text{ for some } k \geq 1. \quad \square$$

A different generalization of Vizing's Theorem to multigraphs has been given by Ore [157]. In order to state his result, we define the <u>enlarged valency</u> $\rho^*(v)$ of a vertex v by

$$\rho^*(v) = \rho(v) + \max_w \mu(vw),$$

where $\rho(v)$ is the usual valency of v, and where the maximum edge-multiplicity is taken over all those vertices w which are adjacent to v. Ore's result now takes the following form:

Theorem 5.4. If M is a multigraph with maximum enlarged valency ρ^*, then $\chi'(M) \leq \rho^*$. $\quad \square$

Shannon's Theorem

We conclude this chapter by showing how Vizing's Theorem for multigraphs (Theorem 5.2) can be used to prove Shannon's result.

Theorem 5.5 (Shannon). If M is a multigraph with maximum valency ρ, then $\chi'(M) \leq [\frac{3}{2}\rho]$.

Proof. Let M be a multigraph for which $\chi'(M) = k$, where $k > [\frac{3}{2}\rho]$. By removing sufficiently many edges from M (if necessary), we may assume that $\chi'(M-e) = k-1$, for each edge e of M. It follows from Theorem 5.2 that $k \leq \rho+\mu$, where μ is the maximum multiplicity of M, and so there must exist vertices v and w for which $\mu(vw) \geq k-\rho$.

We now colour all the edges of M except one of the edges joining v and w; since $\chi'(M-e) = k-1$, this colouring can be done with $k-1$ colours. Now the number of colours missing from v or w (or both) cannot exceed $(k-1) - (\mu-1)$, which in turn cannot exceed ρ, since $k \leq \rho+\mu$. But the number of colours missing from v is at least $(k-1) - (\rho-1) = k-\rho$, and similarly the number of colours missing from w is at least $k-\rho$. It follows that the number of colours missing from *both* v and w is at least $2(k-\rho) - \rho$, which is positive since $k > [\frac{3}{2}\rho]$. By assigning one of these

missing colours to the uncoloured edge joining v and w, we have coloured
all of the edges of M using only k-1 colours, thereby contradicting the
fact that $\chi'(M) = k$. This contradiction establishes the theorem. ☐

Exercises.

5a Verify the statement of Vizing's Theorem when G is (i) a complete
    graph, (ii) a bipartite graph, (iii) a cubic map, (iv) the Petersen
    graph.

5b For each value of ρ find (i) a multigraph for which Vizing's bound
    (Theorem 5.2) is better than Shannon's bound (Theorem 5.5); (ii) a
    multigraph for which Shannon's bound is better than Vizing's bound.

5c We observed in Exercise 3e that if $\omega(G)$ is the clique number of a
    planar graph, then the chromatic number $\chi(G)$ satisfies the inequality
    $\chi(G) \le \omega(G) + 1$.
    (i)   Use Vizing's Theorem to show that the same inequality holds when-
          ever G is a line-graph.
    (ii) By constructing a k-chromatic graph with clique number not
          exceeding k-2, show that the above inequality is false in
          general.   (House [117].)

5d Use the result of Exercise 4k(i) to obtain the analogue of Vizing's
    Theorem for an infinite graph with maximum valency ρ (finite). Find a
    corresponding analogue for a multigraph with maximum valency ρ and
    maximum multiplicity μ.

5e Let M be a multigraph with at least six vertices, and with maximum
    valency ρ. Prove that $\chi'(M) \le [\frac{1}{4}(5\rho+2)]$.

                                    (Andersen [3,4], Gol'dberg [84].)

5f Let M be a multigraph with maximum valency ρ. Prove that
    (i)   if M does not contain the Shannon multigraph with maximum valency
          ρ as an induced submultigraph, then $\chi'(M) < [\frac{3}{2}\rho]$; (Vizing [194].)
    (ii) if M does not contain the Shannon multigraph with maximum valency
          τ as an induced submultigraph (where $\tau \ge \max\{4,\frac{1}{2}\rho\}$), then
          $\chi'(M) \le [\frac{3}{2}\rho] - [\rho/\tau]$.   (Bosák [29].)
    (The result in part (ii) generalizes earlier results of Berge [17] and
    Fiamčik and Jukovič [67].)

5g Let M be a multigraph with maximum valency ρ, and let

$\phi(M)$ = $\max\limits_{H} \left\{ \dfrac{m(H)}{\lceil \frac{1}{2} n(H) \rceil} \right\}$, where the maximum is taken over all submulti-

graphs H of M, and $n(H)$ and $m(H)$ denote the number of vertices
and edges of H. Prove that

(i) $\phi(M) \leq \chi'(M)$;

(ii) if $\chi'(M) > \frac{3}{8}(3\rho+2)$, then $\phi(M) = \chi'(M)$.   (Gol'dberg [87].)

5h Let M be a multigraph with maximum valency ρ, and let
$\psi(M)$ = $\max \{ \lceil \frac{1}{2}(\rho(v) + \rho(w) + \rho(z)) \rceil \}$, where the maximum is taken over
all paths $v \to w \to z$ of length two in M. Prove that
$\chi'(M) \leq \max \{\rho, \psi(M)\}$.   (Ore [157].)

5i A multigraph is called a <u>ring-multigraph</u> if its underlying graph is a
circuit graph. Prove that if M is a ring-multigraph with n vertices,
m edges and maximum valency ρ, then $\chi'(M) = \max \{\rho, 2m/(n-1)\}$.

6 The classification problem

In this chapter we introduce the problem which will be our main concern in
this part of the book - the so-called 'Classification Problem'. We recall
from Vizing's Theorem (Theorem 5.1) that the chromatic index of any graph
with maximum valency ρ must be equal either to ρ or $\rho+1$, and this
immediately gives us a simple way of classifying graphs into two classes,
according to their chromatic index. More precisely, if G is a graph with
maximum valency ρ, we say that G is of class one if $\chi'(G) = \rho$, and that
G is of class two if $\chi'(G) = \rho + 1$. The Classification Problem is the
problem of deciding which graphs are of class one, and which are of class
two.

There is also a corresponding problem for multigraphs. By Vizing's
Theorem for multigraphs (Theorem 5.2), the chromatic index of any multigraph
with maximum valency ρ must be equal to one of the numbers ρ, $\rho+1$, ... ,
$\rho+\mu$, and this immediately gives us a simple way of classifying multigraphs
in a manner analogous to that used for graphs. In this book we shall be
concentrating on the Classification Problem for graphs, although we shall
occasionally refer to the more general problem.

We have already made a start on the Classification Problem for graphs. In
Chapter 4 we saw that circuit graphs and complete graphs are of class one if
the number of vertices is even, and of class two if the number of vertices is
odd. We also proved König's Theorem (Theorem 4.3) which states that all
bipartite graphs are of class one, and we saw in Exercise 4b that the
Petersen graph is of class two. However, the general problem of classifying
graphs as class one or class two remains far from solved. The importance
and difficulty of this problem become apparent when we realize that its
solution would immediately imply the four-colour theorem; this follows from
Tait's result (Theorem 4.4), that the four-colour theorem is equivalent to
the statement that every cubic map is of class one.

Although the general problem of deciding which graphs belong to which
class is unsolved, it certainly seems to be the case that graphs of class two

are relatively scarce. For example, if we look at the 143 connected graphs with at most six vertices, then we can find only eight of class two (see Figure 6.1). A more general result of this kind is due to Erdös and Wilson

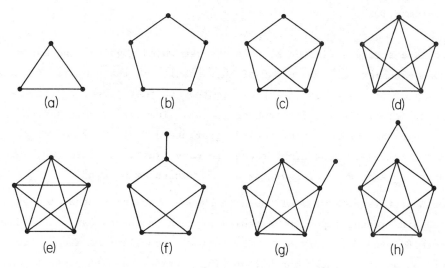

Fig. 6.1

[64], who have proved that almost all graphs are of class one, in the sense that if P(n) is the probability that a random graph on n vertices is of class one, then P(n) → 1 as n → ∞. However, no progress has been made on the more difficult problem of deciding which class contains almost all graphs with a given maximum valency ρ; even for ρ=3 this is unknown.

It seems natural to expect that the more edges a graph has, the more likely it is to be of class two. This idea is made precise in the following result, which gives a sufficient condition for a graph to be of class two. This elementary result was proved by Beineke and Wilson [15], but is implicit in the work of Vizing [194]; note that it applies only when n is odd.

Theorem 6.1. Let G be a graph with n vertices, m edges, and with maximum valency ρ; then G is of class two if $m > \rho[\frac{1}{2}n]$.

Proof. If G is of class one, then any ρ-colouring of the edges of G partitions the set of edges into at most ρ independent subsets. But the number of edges in each independent subset cannot exceed $[\frac{1}{2}n]$, since otherwise two of these edges would be adjacent. It follows that $m \leq \rho[\frac{1}{2}n]$, giving the required contradiction. □

Corollary 6.2. If G is a graph of odd order with maximum valency ρ and total deficiency less than ρ, then G is of class two.

Proof. If G has n vertices and m edges, then the total deficiency is $n\rho - 2m$. But if $n\rho - 2m < \rho$, then $m > \frac{1}{2}(n-1)\rho = \rho[\frac{1}{2}n]$, and the result follows from Theorem 6.1. □

There is a stronger version of this theorem which involves the edge-independence number α. The proof of this result - that G is of class two if $m > \alpha\rho$ - is almost identical to the one just given, and the reader is asked to supply the details in Exercise 6c. The reader should also note that although the condition $m > \rho[\frac{1}{2}n]$ guarantees that G is of class two, it is not a necessary condition; there are many graphs of class two, such as the graphs P and $P-v$ of Exercise 4b, for which $m \le \rho[\frac{1}{2}n]$.

Throughout this book we shall need to have a stock of graphs of class two. We can obtain several important examples of such graphs by using the results of Theorem 6.1 and Corollary 6.2. The following corollaries are due to Vizing [194] and to Beineke and Wilson [15]; further results of this kind may be found in Exercise 6a.

Corollary 6.3. If G is a regular graph with an odd number of vertices, then G is of class two.

Proof. Since the total deficiency of G is zero, the result follows immediately from Corollary 6.2. □

We can construct examples of such regular graphs by using the well-known fact that the complete graph K_{2s+1} can be split into s edge-disjoint Hamiltonian circuits (see, for example, [99, p.89]). If we let G be a graph obtained by superimposing any number of these circuits, then G must be of class two. To show how this construction works in practice, let us take successively one, two, three, and four edge-disjoint Hamiltonian circuits in K_9; we then obtain the four graphs of class two shown in Figure 6.2.

In fact, the result of Corollary 6.3 can be somewhat strengthened. If we have a regular graph with an odd number of vertices (other than an odd circuit graph), then we can always remove some of its edges, and still be

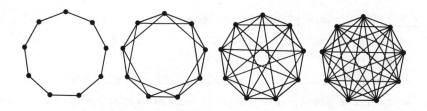

Fig. 6.2

left with a graph of class two. A precise statement of this result is as follows:

Corollary 6.4. If H is a regular ρ-valent graph with an odd number of vertices, and if G is any graph obtained from H by deleting not more than $\frac{1}{2}\rho-1$ edges, then G is of class two.
Proof. Since the total deficiency of G is at most $\rho-2$, the result follows immediately from Corollary 6.2. ☐

This result is in general best possible, as can be seen by considering the last three graphs of Figure 6.2. By Corollary 6.4, we can remove any one, two or three edges (respectively) from them, and still obtain a graph of class two; however, as we ask you to show in Exercise 6e, the removal of more edges always results in a graph of class one.

Another very useful method for constructing graphs of class two is to take a regular graph and insert a vertex into one of its edges (see Figure 2.3). The precise formulation of this result is as follows:

Corollary 6.5. If H is a regular graph with an even number of vertices, and if G is any graph obtained from H by inserting a new vertex into one edge of H, then G is of class two.
Proof. If H has n vertices, and is regular with valency ρ, then G has n+1 vertices and $\frac{1}{2}n\rho+1$ edges. The result follows from Theorem 6.1. ☐

We conclude this chapter on the Classification Problem by stating a remarkable result of Vizing [195]; it concerns the classification of planar graphs with sufficiently large maximum valency.

Theorem 6.6. If G is a planar graph whose maximum valency is at least 8, then G is of class one. ☐

Although the statement of this surprising theorem fits nicely into the present chapter, its proof, like those of several other results of a similar nature, will have to wait until after we have developed the machinery of critical graphs.

Exercises.

6a Prove the following corollaries of Theorem 6.1:

   (i)   if H is any graph with an odd number of vertices, one of which has valency t and the rest of which have valency ρ (where $\rho \geq t$), and if G is any graph obtained from H by deleting not more than $\frac{1}{2}t-1$ edges, then G is of class two.

   (ii) if G is any graph obtained by taking an odd circuit C_{2s+1} and adding to it not more than 2s-2 sets of s independent edges, then G is of class two.   (Beineke and Wilson [15].)

6b Prove that if G is a regular graph which contains a cut-vertex, then G is of class two.   (Vizing [194].)

6c Let G be a graph with m edges, maximum valency ρ, and edge-independence number α, where $m > \rho\alpha$; show that G is of class two.

6d (i)   Use the results of Corollaries 6.3 and 6.5 to prove that if ρ is given, then there exist graphs of class two whose maximum valency is ρ.

   (ii) Prove further that if ρ and g are given, then there exist graphs of class two whose maximum valency is ρ and whose girth is at least g.   (Vizing [194].)

6e Prove that if we remove more than one, two, or three edges (respectively) from the last three graphs of Figure 6.2, we obtain graphs of class one. Can this result be generalized to graphs with a larger number of vertices?

6f By applying Corollary 6.5 to the platonic graphs, construct planar

40

graphs of class two with maximum valencies 3, 4 and 5.

6g It can be shown that if S is a 1-factor of the complete graph K_{2s},
then $K_{2s}-S$ may be split into s-1 edge- disjoint Hamiltonian circuits.
Use this fact to prove that $K_{2s}-S$ is of class one.

6h Let G be a graph of class one, in which the valency of each vertex is
either ρ or 1. For any ρ-colouring of the edges of G, let f_i be
the number of terminal edges with colour i, for i = 1,2,...,ρ. Prove
that all the f_i's have the same parity.   (Izbicki [122].)

6i (i)   If G is a regular graph of class one with an even number of
           edges, prove that its line-graph L(G) is also of class one.
                                                        (Jaeger [126].)

    (ii)   If G is a cubic graph with an even number of edges, and if L(G)
           is of class one, prove that G is of class one.
                                          (Seymour [165], Kotzig [137].)

    (iii) Show that the graphs $L(K_7)$, $L(L(K_7))$, $L(L(L(K_7)))$, ... are all of
           class two.

6j Prove that every regular graph with 4, 6 or 8 vertices is of class one.

7 Petersen-type graphs

In the previous chapter we introduced the Classification Problem which is concerned with the classification of graphs into two classes according to whether their chromatic index is equal to, or one more than, their maximum valency. Much of the interest in this problem has arisen out of the study of regular graphs, and it is these graphs to which we shall be devoting our attention in this chapter and the next. We already know from Corollary 6.3 that if G is a regular graph with an odd number of vertices, then G is necessarily of class two. The difficulty arises when we consider regular graphs with an even number of vertices, since some of them (such as the even complete graphs and the platonic graphs) are of class one, whereas others (such as the Petersen graph) are of class two. Unfortunately, very few general results concerning these graphs are known, although the even-order regular graphs of class two seem to be relatively hard to find.

Most of the even-order regular graphs of class two that are known turn out to be related in some way to the Petersen graph. In this chapter we shall describe some of the various ways in which the Petersen graph has been generalized, and we shall determine the corresponding chromatic index in each case; we shall find that we sometimes end up with graphs of class one, and sometimes with graphs of class two, and that the graphs we obtain are of considerable interest both here and elsewhere. A more general study of regular graphs and their edge-colouring properties will be given in Chapter 8.

Before starting our discussion it will be convenient to present four different drawings of the Petersen graph (see Figure 7.1).

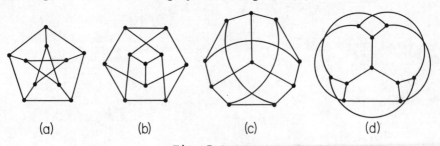

    (a)   (b)   (c)   (d)

Fig. 7.1

Note that the first three of these diagrams have bounding circuits with five, six and nine edges, respectively.

The Generalized Petersen Graphs

The Petersen graph may be obtained by taking an outer circuit with five vertices, five spokes incident to the vertices of this 5-circuit, and an inner 5-circuit attached by joining its vertices to every *second* spoke (see Figure 7.1(a)). M.E. Watkins [198] has defined the <u>generalized Petersen graph</u> $P(n,k)$, which consists of an outer n-circuit, n spokes incident to the vertices of this n-circuit, and an inner n-circuit attached by joining its vertices to every *k'th* spoke. Thus the Petersen graph is the graph $P(5,2)$, and $P(9,2)$ is the graph shown in Figure 7.2. (We shall be meeting $P(9,2)$ again in Chapter 18.) Note that $P(n,k)$ is isomorphic to $P(n,n-k)$.

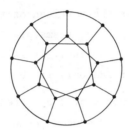

Fig. 7.2

Watkins conjectured that, with the single exception of the Petersen graph itself, all of these graphs are of class one. He later succeeded in settling several cases of this conjecture himself [198], and further results of this nature were obtained by Kaiser and Walther [131]. In 1973, the conjecture was finally settled by Castagna and Prins [42], who proved the following result:

Theorem 7.1. All of the generalized Petersen graphs are of class one, except the Petersen graph. □

The Odd Graphs

For each integer $k \geq 2$, the <u>odd graph</u> O_k is the graph obtained by taking as vertices each of the (k-1)-subsets of $\{1,2,\ldots,2k-1\}$, and joining two of these vertices by an edge whenever the corresponding subsets are disjoint. (These graphs were first studied by Balaban, Françasiu and Banica in a chemical context [7], under the name "k-valent halved combination graphs".) It is easy to verify that O_2 is simply a triangle, and that O_3 is the Petersen graph. More generally, it is clear that the graph O_k is always regular of valency k, and that O_k is of odd order if and only if k is a power of 2.

Interest in the edge-colouring properties of the odd graphs arose out of the following problem of Biggs [19]:

'The eleven footballers of Croam play five-a-side matches with the eleventh man as referee, and each possible choice of teams and referees play exactly one match. Is it possible to schedule all of the 1386 games in such a way that each individual team plays its six games on six different weekdays?'

By taking the various teams as the vertices of the odd graph O_6, joining two vertices whenever the corresponding teams play each other, and letting the six weekdays be represented by colours, it is easily seen that the scheduling of the games is possible if and only if O_6 is of class one.

That this is indeed the case was proved by Meredith and Lloyd [152], who have investigated the edge-colouring properties of several of the odd graphs. We already know that O_3 (the Petersen graph) is of class two, and that the graphs O_2, O_4 and O_8 are also of class two, since they are regular graphs with an odd number of vertices (see Corollary 6.3). Meredith and Lloyd proved that O_5 and O_6 are of class one, by first establishing that O_5 is an edge-disjoint union of two Hamiltonian circuits and a 1-factor, and O_6 is an edge-disjoint union of three Hamiltonian circuits. They also showed [151] that O_7 is a Hamiltonian graph, and the corresponding result for O_8 was later proved by Mather [144].

It seems likely that all of the odd graphs O_k (k ≥ 4) are Hamiltonian, that the graphs O_{2k} (k ≥ 2) may be split up into k edge-disjoint Hamiltonian circuits, and that the graphs O_{2k+1} (k ≥ 2) may be split up into k edge-disjoint Hamiltonian circuits, and a 1-factor. If these results are true they would immediately settle the 'odd graph conjecture', which we state formally as follows:

Odd Graph Conjecture: O_k is of class one unless $k = 3$ or $k = 2^r$ for some integer r.

Meredith's Graphs

Let $k \geq 2$, and let r be the integer nearest to $\frac{k}{3}$ (so that $k = 3r$ or $3r \pm 1$). G.H.J. Meredith [150] has defined a graph G_k which is obtained by taking the Petersen graph and replacing each vertex by a copy of the complete bipartite graph $K_{k,k-1}$, joined up as shown in Figure 7.3, which represents G_4.

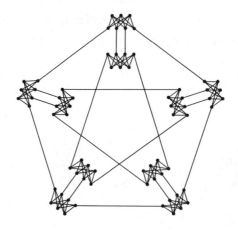

Fig. 7.3

Meredith proved that the graph G_k has $20k - 10$ vertices, is regular of valency k, and is always non-Hamiltonian and k-connected. He also investigated the chromatic index of these graphs, and proved that G_k is of class one whenever r is even, and of class two whenever r is odd.

Biggs' "Remarkable Graphs"

In [21], N.L. Biggs describes three 'remarkable graphs' which have a high degree of symmetry. These graphs may be described conveniently by means of the following diagram (see Figure 7.4), which may be interpreted in the following way. The first graph is obtained by taking five copies of the line, joining one end of each of these edges successively to the vertices of a 5-circuit, and the other end successively to every second vertex of another 5-circuit; as we might expect, this gives the Petersen graph. The second graph

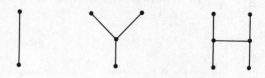

Fig. 7.4

is obtained by taking seven copies of the Y-shaped diagram, and joining the
corresponding pendant vertices successively to every vertex of a 7-circuit,
every second vertex of a second 7-circuit, and every fourth vertex of a third
7-circuit; this gives a graph with 28 vertices, known as the _Coxeter graph_
(see Figure 7.5(a)). The third graph is obtained by taking 17 copies of

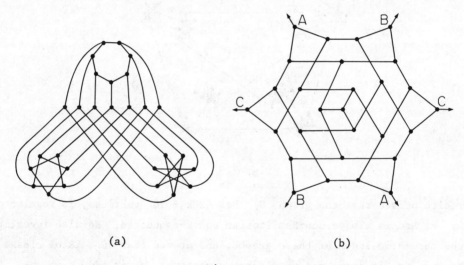

(a)   (b)

Fig. 7.5

the H-shaped diagram, and joining the corresponding pendant vertices successively to every vertex of a 17-circuit, every second vertex of a second 17-
circuit, every fourth vertex of a third 17-circuit, and every eighth vertex
of a fourth 17-circuit; this gives a graph with 102 vertices, discovered
independently by Biggs and Smith and by J.H. Conway.

It turns out that all three of these graphs are distance-transitive (that
is, if v, w and v', w' are any two pairs of vertices with the property
that the length of the shortest path from v to w is the same as that from

46

v' to w', then there is an automorphism of the graph which sends v to v' and w to w'). This means that they can be redrawn with a vertex in the centre, neighbouring vertices around it, neighbouring vertices around them, and so on; the corresponding drawings for the Petersen graph and the Coxeter graph are given in Figures 7.1(b) and 7.5(b). In fact, these three graphs, together with K_4, are the *only* trivalent distance-transitive graphs in which the automorphism group acts primitively on the vertices (see [21]).

In view of all this, one might expect that all three of these graphs would have the same chromatic index, but surprisingly this is not the case – although the Petersen graph is of class two, the Coxeter graph and the graph on 102 vertices are both graphs of class one.

'Snarks'

In recent years, much attention has been paid to the search for bridgeless cubic graphs with chromatic index four. Part of the interest was in trying to prove or disprove the existence of a planar graph of this type, since this would settle the four-colour problem (see Theorem 4.4). Now that the four-colour theorem has been proved, and cubic maps of class two are known not to exist (Corollary 4.5), the search is directed towards non-planar cubic graphs of class two. Because such graphs are so difficult to find, Martin Gardner, in a delightful popular article on the subject [82], has christened them 'Snarks' (a name borrowed from Lewis Carroll's 'The Hunting of the Snark'). More precisely, a snark is a cubic graph G of class two whose girth is at least five. To avoid trivial cases, we shall further assume that G does not contain three edges whose deletion results in a disconnected graph, each of whose components is non-trivial.

The smallest of all snarks is the Petersen graph, which dates from 1898. About fifty years later two further snarks were uncovered – the Blanuša snark with 18 vertices, discovered in 1946 [25], and the Descartes snark with 210 vertices, discovered in 1948 [49]. The fourth snark to be found was the Szekeres snark with 50 vertices, discovered in 1973 [173]. Until 1973, these were the only snarks known.

The art of snark-hunting underwent a dramatic change in 1975, when R. Isaacs published an important article [119] in which he described two infinite families of snarks, one of which essentially included the known snarks, and the other of which was completely new. He also discovered one extra snark - the double-star snark, with 30 vertices - which does not belong to either family (see Figure 7.6).

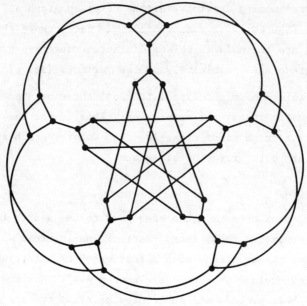

Fig. 7.6

Isaacs' first family is obtained by taking graphs of class two, and 'hooking them together' in such a way as to give new graphs of class two. Although this construction can be described in general terms, we shall illustrate it only in the case of two Petersen graphs, drawn as in Figure 7.1(c). In order to combine them, we remove any two non-adjacent edges from one of them, and two adjacent vertices (and the five incident edges) from the other, and hook them together as shown in Figure 7.7; the result in this case is in fact the Blanuša snark on 18 vertices. Further snarks may be obtained by inserting

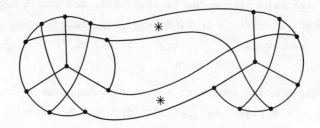

Fig. 7.7

arbitrary graphs at the places marked (*), or by crossing either or both of
the corresponding pairs of edges. Using variations of this kind, Isaacs was
able to re-create the Szekeres snark (obtained by hooking together six Petersen graphs) and the Descartes snark (obtained by hooking together several
Petersen graphs and inserting 9-circuits). The reader is referred to Isaacs'
article [119] for further details.

Isaacs' second family is easier to describe. The first member of the
family is obtained by replacing the central vertex of the Petersen graph (in
the drawing of Figure 7.1(d)) by a triangle. The rest of the family is then
obtained by replacing the three large 'petals' by five, seven, nine,...
petals; this gives a series of snarks with 12, 20, 28, 36,... vertices, the
first three of which are shown in Figure 7.8. (Note that the first of these

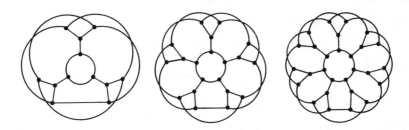

Fig. 7.8

is not really a snark since it contains a triangle.) Because of their general appearance, these snarks are sometimes called 'flower snarks'.

More recently, a new double family of snarks has been found by Loupekhine;
the reader is referred to [120] for further details. In addition, Fiorini
has used a computer-generated list of Bussemaker *et al.* [36] to show that
there are no snarks of order 12 or 14. On the other hand, it can be shown
that snarks exist for any even order greater than 16. Whether there are any
snarks whose order is exactly 16 is, however, still unsolved.

Tutte's Conjecture

We conclude this chapter with a conjecture of Tutte. The reader may well
have been asking why we have been so involved in this chapter with so many
generalizations of the Petersen graph, especially when several of them may

seem at first sight to be somewhat special. The reason is that every known bridgeless cubic graph of class two does, in fact, 'contain' the Petersen graph, and Tutte [183] has conjectured that this is always the case. For results relating to this conjecture, the reader is referred to [161,165].

Tutte's Conjecture: Every bridgeless cubic graph of class two contains a subgraph homeomorphic to the Petersen graph.

Exercises

7a (i) Find explicit 3-colourings of the edges of (a) the generalized Petersen graph $P(7,3)$, (b) the Coxeter graph.

    (ii) Find explicit 4-colourings of the edges of (a) the Meredith graph G_3, (b) the double-star snark.

7b Prove that every 3-colouring of the edges of the graph $P(9,2)$ partitions the edge-set into edges of the same colour.

7c Prove that the odd graph O_4 is Hamiltonian.

7d Prove that the Meredith graph G_4 is 4-connected and non-Hamiltonian.

7e Prove that the Coxeter graph is non-Hamiltonian.   (Tutte [181].)

7f Verify Tutte's Conjecture for each of the graphs of class two mentioned in this chapter.

7g Prove that Tutte's Conjecture is equivalent to the statement that every bridgeless cubic graph of class two contains a subgraph contractible to the Petersen graph.

8 Regular graphs

In the previous chapter we looked at the chromatic index of various regular graphs derived from the Petersen graph. Our aim in this chapter is to look at some of the more general properties of regular graphs, especially those of class one. As in the last chapter, we shall sometimes be content to state our results without proof, referring the reader to the original papers for further details. Group-theoretical definitions not included here may be found in any of the standard textbooks on group theory.

Permutation Graphs

We start by looking at the concept of a permutation graph, which was first defined by Chartrand and Harary [45], and which generalizes the construction of the generalized Petersen graphs $P(n,k)$. If G is a graph whose vertices are labelled $1,2,\ldots,n$, and if G' is another copy of G, but with the same vertices labelled $\alpha(1),\alpha(2),\ldots,\alpha(n)$ (where α is a permutation of the set $\{1,2,\ldots,n\}$), then the <u>permutation graph</u> $P_\alpha(G)$ is the graph formed by taking the disjoint union of G and G' and adding the edges joining i to $\alpha(i)$, for each i. For example, if α is the permutation $\begin{pmatrix} 1 & 2 & 3 & 4 & 5 \\ 1 & 3 & 5 & 2 & 4 \end{pmatrix}$, then the Petersen graph is the graph $P_\alpha(C_5)$.

In 1969, Chartrand and Frechen [44] asked whether it would be possible to characterize those permutations α for which $P_\alpha(C_n)$ is of class one. This problem was investigated by Zelinka [205], who obtained the following interesting results:

<u>Theorem 8.1.</u> Let S_n be the symmetric group on the set $\{1,2,\ldots,n\}$ (where $n \geq 3$), and let M_n be the set of those permutations α for which $P_\alpha(C_n)$ is of class one. Then
(i)   if n is even, then $M_n = S_n$;
(ii) if n is odd, then either $M_n = S_n$, or M_n is not a group. $\square$

Zelinka also obtained several further results in the case where n is odd and M_n is not a group. In particular, he proved that

(i)   if α is any transposition, then $\alpha \in M_n$;

(ii) the product of any permutation in M_n with the cycle permutation

$$\begin{pmatrix} 1 & 2 & 3 & \cdots & n \\ 2 & 3 & 4 & \cdots & 1 \end{pmatrix}$$ is also in M_n;

(iii) the product of any permutation in M_n with the 'mirror permutation'

$$\begin{pmatrix} 1 & 2 & 3 & \cdots & n \\ n & n-1 & n-2 & \cdots & 1 \end{pmatrix}$$ is also in M_n.

For further details, the reader is referred to [205].

Some results of Biggs

Group-theoretical considerations of a rather different nature have arisen in connection with N.L. Biggs' work on regular graphs of class one. If G is a regular ρ-valent connected graph of class one, and $(c_1,\ldots,c_\rho)$ is a given colouring of the edges of G, then we can define A to be the group of all those automorphisms of G which preserve the colours of the edges of G. It turns out that there is an interesting interplay between the group A and another group S which we now define. For each $i = 1,2,\ldots,\rho$, we let α_i be the permutation defined by $\alpha_i(v) = w$ if and only if vw is an edge of G coloured c_i, so that each α_i is a product of n disjoint transpositions; we can then define S to be the group generated by the permutations $\alpha_1,\alpha_2,\ldots,\alpha_\rho$. The following simple results have been proved by Biggs [20]:

Theorem 8.2. (i) The permutation group A acts without fixed points on $V(G)$;   (ii) The permutation group S acts transitively on $V(G)$.

Proof. (i) If f is an automorphism in A which fixes the vertex v, then every vertex adjacent to v is also fixed, since f preserves edge-colours. Iterating this, we see that f fixes every vertex of G (since G is a connected graph), and must therefore be the identity permutation. (ii) If v and w are any two vertices of G joined by a path whose edges are coloured successively with colours $c_{i_1},c_{i_2}, \ldots ,c_{i_s}$, then the corresponding permutation $\alpha_{i_1}\alpha_{i_2}\ldots\alpha_{i_s}$ in S takes v to w. The result follows. $\square$

Using these results, Biggs showed that the group A is in fact the centralizer of S in the symmetric group of all permutations of $V(G)$, and that A is a regular permutation group (that is, acts transitively and without fixed points) if and only if S is also a regular permutation group.

The Enumeration of 3-colourings

In the same paper, Biggs also obtained an expression for the number of ways of colouring the edges of a regular ρ-valent graph G of class one, in terms of the determinant of a related matrix. If f is any function from $E(G)$ to the set $\{1,2,\ldots,\rho\}$, and if, for each $i = 1,2,\ldots,\rho$, the matrix F_i is defined by $(F_i)_{jk}$ equals 1, if $v_j v_k$ is an edge of G with $f(v_j v_k) = i$, and 0 otherwise, then $F_1 + F_2 + \ldots + F_\rho = A(G)$, the adjacency matrix of G. We then obtain the following theorem:

Theorem 8.3. The following statements are equivalent:
(i) f is an edge-colouring of G;
(ii)   for each $i = 1,2,\ldots,\rho$, F_i is a permutation matrix;
(iii) $\det(F_1 F_2 \ldots F_\rho) \neq 0$.
Proof.
(i) $\rightarrow$ (ii): If f is an edge-colouring of G, then each colour appears exactly once at each vertex, and so F_i must be a permutation matrix.
(ii) $\rightarrow$ (iii): If each F_i is a permutation matrix, then so is $F_1 F_2 \ldots F_\rho$; it follows that $F_1 F_2 \ldots F_\rho$ is invertible, and hence that $\det(F_1 \ldots F_\rho) \neq 0$.
(iii) $\rightarrow$ (i): If f is not an edge-colouring of G, then there exist a vertex v_i and a colour c_j with the property that none of the edges incident to v_i has colour c_j, so that F_j has a row and column consisting entirely of zeros. It follows that $\det(F_j) = 0$, and hence that $\det(F_1 F_2 \ldots F_\rho) = 0$, contradicting (iii). $\square$

Corollary 8.4. If G is a regular ρ-valent graph of class one, with m edges, then the number of ways of colouring the edges of G with ρ given colours is precisely

$$(-1)^m \sum_f \det(F_1 F_2 \ldots F_\rho),$$

where the sum is taken over all functions f from $E(G)$ to the set $\{1,2,\ldots,\rho\}$.

53

Proof. If f is not an edge-colouring of G, then $\det(F_1 F_2 \ldots F_\rho) = 0$. If f is an edge-colouring of G, then each matrix F_i is the product of $\tfrac{1}{2}n$ transpositions (where n is the number of vertices of G). It follows that $\det(F_i) = (-1)^{\frac{1}{2}n}$, and hence that $\det(F_1 F_2 \ldots F_\rho) = (-1)^{\frac{1}{2}n\rho} = (-1)^m$. The result now follows immediately. $\square$

If G is a cubic map, then alternative methods of calculating the number of edge-colourings of G have been given by D.E. Scheim [163] and R. Penrose [159]. Penrose's method involves the use of formulas in tensor calculus, and will be described in the next chapter. In order to describe Scheim's result, we first define, for any graph H, a $(0,1,-1)$ matrix $M(H) = (m_{ij})$; this matrix is obtained by giving each edge of H an arbitrary orientation, and defining for each vertex v_i and edge e_j,

$$m_{ij} = \begin{cases} 0, & \text{if } e_j \text{ is not incident to } v_i \\ 1, & \text{if } e_j \text{ is oriented out of } v_i \\ -1, & \text{if } e_j \text{ is oriented into } v_i. \end{cases}$$

We then let $M^*(H)$ be the matrix obtained by writing each row of $M(H)$ twice; for example, if $H = K_3$

and $M(H) = \begin{pmatrix} 1 & 0 & -1 \\ -1 & 1 & 0 \\ 0 & -1 & 1 \end{pmatrix}$, then $M^*(H) = \begin{pmatrix} 1 & 0 & -1 \\ 1 & 0 & -1 \\ -1 & 1 & 0 \\ -1 & 1 & 0 \\ 0 & -1 & 1 \\ 0 & -1 & 1 \end{pmatrix}$.

Scheim's result may now be stated as follows:

Theorem 8.5. If G is a cubic map with m edges, and if $L(G)$ is the line-graph of G, then the number of 3-colourings of the edges of G is equal to

$$2^{-m} |\text{permanent of } M^*(L(G))|. \quad \square$$

Some results of Szekeres

We conclude this chapter by turning to the work of Szekeres on cubic graphs of class one. If we take any particular 3-colouring of such a graph G with colours α, β, γ, then the two-coloured subgraphs $H(\alpha,\beta)$, $H(\alpha,\gamma)$ and

54

H(β,γ) (defined in Chapter 5) are comprised of circuits of G, which we call
basic circuits of G; note that each edge of G lies on exactly two of these
basic circuits. If we now give an arbitrary orientation to each of these
circuits, then we can define the orientation index of G to be equal to the
number of edges of G for which the two orientations disagree. Using this
notation, Szekeres obtained the following result [170]:

Theorem 8.6. If G is a cubic graph of class one, then the orientation
index of G has the same parity as the number of basic circuits of G, and
is therefore independent of the choice of orientation given to the basic
circuits.   ☐

   The importance of Szekeres's work is that it throws useful light on the
following conjecture of Grünbaum [90]:

Grünbaum's Conjecture (First form). A cubic graph is of class one if it can
be embedded in an orientable closed surface in such a way as to divide the
surface into simply-connected regions bounded by circuits no two of which
have more than one edge in common.

   Motivated by this conjecture, Szekeres [171] defined a polyhedral decomposition of a cubic graph G to be a set of circuits in G with the
property that each edge of G lies in exactly two of them; if all of these
circuits have even length, then the decomposition is called even. By our
above discussion, it is clear that G has an even polyhedral decomposition
if G is of class one; surprisingly, the converse result also turns out to
be true:

Theorem 8.7. A cubic graph is of class one if and only if it has an even
polyhedral decomposition.   ☐

   To see the relevance of these results to the four-colour theorem and to
Grünbaum's Conjecture, Szekeres defined a polyhedral decomposition to be
proper if no two of the circuits have more than one common edge, and coherent
if each circuit can be given an orientation in such a way that each edge of
G is traversed in opposite directions in the two circuits which contain it.

55

Stated in these terms, the close relationship between the four-colour theorem and Grünbaum's Conjecture becomes clearly apparent:

The Four-Colour Theorem: Every cubic graph which has a proper coherent polyhedral decomposition of genus 0 has an even polyhedral decomposition.

Grünbaum's Conjecture (Second Form): Every cubic graph which has a proper coherent polyhedral decomposition has an even polyhedral decomposition.

Exercises.

8a Find a permutation α of the set $\{1,2,\ldots,n\}$ such that $P_\alpha(C_n) = P(n,k)$.

8b Find conditions sufficient to ensure that the group S mentioned in Theorem 7.2 is a symmetric or an alternating group.   (Biggs [20].)

8c Show that the group of automorphisms of the Coxeter graph (Figure 7.5) acts transitively on the set of 1-factors. Hence, or otherwise, show that the Coxeter graph has exactly 84 1-factors and 56 distinct edge-colourings with three colours.   (Biggs [21].)

8d Verify the results of Corollary 8.4 and Theorem 8.5 when G is the complete graph K_4.

8e (i) Show that the Petersen graph has no coherent polyhedral decomposition, and is therefore not a counterexample to Grünbaum's Conjecture.

    (ii) Give an example to show that the converse of Grünbaum's Conjecture is false.

8f Let D be a directed graph, and let C be a cutset which disconnects D into sub-digraphs A and B ; let s be the number of arcs of C directed from A to B, and t be the number directed from B to A. Assuming, without loss of generality, that $s \geq t$, we define the flow-ratio of C to be $\frac{s}{t}$. Prove that a cubic graph G is of class one if and only if the edges of G can be oriented in such a way that the flow-ratio of each cutset of G does not exceed 3.   (Minty [153].)

56

9 Some applications

The object of this chapter is to describe the inter-relationships which exist between the theory of edge-colourings and such diverse fields as network analysis, scheduling problems and statistics. We shall also show how results in tensor calculus have been used to yield information about edge-colourings of planar graphs.

Electrical Networks

One of the most important contributions to the theory of edge-colourings was a paper on electrical networks, written by C.E. Shannon in 1949 [166]; in this paper he considered the following problem:
'Suppose we have an electrical unit, such as a relay panel, and there are a number of relays, switches, and other devices A,B,...,E to be inter-connected; the connecting wires are first formed into a cable, with the wires to be connected to A emerging at one point, those connected to B emerging at another, and so on. In order to distinguish them, it is' necessary that all those wires which emerge from the same point are differently coloured. What is the least number of colours necessary for the whole network?'

In his paper Shannon obtained the following result, which has already been proved earlier in this book (Theorem 5.5); networks which actually need the specified number of colours were described in Exercise 4c.

Theorem 9.1. If ρ is the largest number of wires meeting at any point of a network, then the wires can be coloured with $[\frac{3}{2}\rho]$ colours in such a way that no two wires with a common junction have the same colour. $\square$

Scheduling Problems

There are several scheduling problems in Operational Research which can be formulated in graph-theoretical terms, and then solved using edge-colouring methods. We shall consider a few of these problems, some trivial and others somewhat deeper.

Lucas' Schoolgirls Problem: 'Each day 2k schoolgirls take a walk in k pairs; how many walks can they take without any two girls walking together more than once?' The answer to this problem is 2k-1, corresponding to the fact that the chromatic index of the complete graph K_{2k} is 2k-1.

Examination Scheduling: 'At the end of an academic year, each student must be examined orally by each of the professors who have taught him; how many examination periods are needed?' To solve this problem we form the bipartite graph whose vertices are the students and the professors, with edges joining each student to the professors who have taught him. By König's Theorem (Theorem 4.3), the chromatic index of this bipartite graph is equal to its largest valency, and this will be the number of examination periods required.

The General Scheduling Problem: If V is a set of workers, W a set of machines, and U a set of jobs each of which has to be performed on a pre-scribed machine, then we can construct the bipartite multigraph G = (V∪W,U) in which the vertices v and w are joined by p edges if and only if a worker v has to perform a job on a machine w in p periods of time. The scheduling problem is then to partition the edge-set U into the smallest possible number of colour-classes $C_1, \ldots, C_k$, just as in the examination problem above. In practice, however, there are often other restrictions imposed by the problem in hand. For example, if there is only a limited amount of some resource, such as electric power, then it may be important to require the colour-classes to be as nearly equal in size as possible; in this connection, D. de Werra [52] and C.J.H. McDiarmid [145] have proved independently that the partitioning of the edges of any graph can be arranged in such a way that any two colour-classes differ in size by at most 1.

An Experimental Design Problem: Suppose that we wish to test the effect on rats of certain experiments applied in pairs. If $X = \{x_1, \ldots, x_n\}$ is the set of experiments, and if a_{ij} is the number of times we wish to perform the pair of experiments x_i and x_j, then we can form the bipartite multigraph of order 2n, in which the vertices in each set correspond to the experiments, and where the vertices corresponding to x_i and x_j are joined by a_{ij} edges. If each experiment lasts exactly one day, then the chromatic index of this bipartite multigraph is the minimum number of days required to complete the experiments. Note that, by the above-mentioned result of de Werra and McDiarmid, we can always arrange the experiments in such a way that the number performed on each day is as nearly constant as possible.

Further Results: Problems of the type considered by de Werra and McDiarmid have also been considered by Folkman and Fulkerson [79], who posed the following general question: 'suppose that we are given a finite sequence of positive integers $p_1, p_2, \ldots, p_k$; under what conditions can the edges of a given graph G be coloured with k colours in such a way that precisely p_i edges of G have colour c_i, for each i?' In this general form, the problem has as yet no solution, although partial results are known (see, for example, [50-52, 145]).

If G is a bipartite graph, we can give the following physical interpretation to this problem. 'A firm has applicants $\{A_1, A_2, \ldots, A_m\}$ for vacancies $\{V_1, V_2, \ldots, V_n\}$; under what conditions can one carry out supervised trials, placing the applicants in the corresponding vacancies for a week at a time, in such a way that the number of trials in the i'th week is p_i?' In connection with this problem, Geller and Hilton [83] have proved the following result:

Theorem 9.2. If G is a bipartite graph, whose valencies are $d_1, d_2, \ldots, d_k (=\rho)$, where $d_i < d_{i+1}$ $(i = 1, 2, \ldots, \rho-1)$, then the edges of G can be coloured with colours $c_1, c_2, \ldots, c_\rho$ in such a way that every edge which is coloured c_{d_i+1} (for each i) is incident to vertices whose valency is at least d_{i+1}. $\square$

The following related problem for complete bipartite graphs remains unsolved:

Conjecture. The edges of $K_{\rho,\rho}$ can always be coloured with ρ colours in such a way that any given set of $\rho-1$ edges are assigned previously-determined colours.

Similar problems, expressed in the framework of matroid theory, have been posed by Welsh [199].

Matrix Algebra

McDiarmid [145] has also used some of the above results in connection with a problem in matrix algebra; in order to state his results, we need some terminology. A <u>rational doubly-substochastic matrix</u> is a matrix of non-negative rational numbers in which each row and column sum does not exceed 1; a <u>subpermutation matrix</u> is a matrix in which each row and column contains at most one entry equal to 1, and all other entries are zero; the <u>size</u> of a matrix with non-negative entries is the sum of its entries. McDiarmid's results may now be stated as follows:

<u>Theorem 9.3.</u> Let M be a rational doubly-substochastic matrix of size E, and let the least common denominator of its elements be p. Then there exist subpermutation matrices $M_1, \ldots, M_p$ of size $[E]$ or $\{E\}$, such that

$$M = \frac{1}{p} \sum_{i=1}^{p} M_i .$$

Conversely, suppose that M is a symmetric matrix of size E, and that $M_1, \ldots, M_p$ are symmetric subpermutation matrices such that

$$M = \frac{1}{q} \sum_{i=1}^{p} M_i , \text{ for some } q.$$

Then there exist symmetric subpermutation matrices $N_1, \ldots, N_p$ of size $[qE/p]$ or $\{qE/p\}$, such that

$$M = \frac{1}{q} \sum_{i=1}^{p} N_i . \quad \square$$

Similar, but less exact, results have also been obtained for infinite matrices. It is worth pointing out that the converse result in Theorem 9.3 is not merely an existence theorem; if the matrices M_i are given, then the N_i can be obtained explicitly by using the Hungarian algorithm (see [27, p.82]).

Latin Squares

Latin squares are frequently encountered in Statistics in connection with the design of experiments. One problem which sometimes arises concerns the conditions under which a given Latin square can be embedded in a larger one. In particular, if we are given a symmetric $k \times k$ Latin square on $\{1, 2, \ldots, n\}$, where $k < n$, we can ask for conditions under which we can embed it in a symmetric $n \times n$ Latin square.

Necessary and sufficient conditions under which this can be done are known. However, as A.J.W. Hilton [113] has pointed out, problems of this kind can sometimes be simplified by reformulating them in terms of edge-colourings. We shall not go into details here, but will be content simply to give an example of an edge-colouring result and its Latin square analogue. The edge-colouring result, due to Hilton, is as follows:

Theorem 9.4(a). If $n > k$, and n is odd, then an edge-colouring of K_k with n colours can be extended to an edge-colouring of K_n with n colours if and only if each colour occurs on at least $k - \frac{1}{2}(n+1)$ edges of K_k; a similar result holds if n is even. $\square$

Its Latin square analogue, due to A. Cruse [47] is as follows:

Theorem 9.4(b). If $n > k$, a symmetric $k \times k$ Latin square on $\{1, 2, \ldots, n\}$ can be extended to a symmetric $n \times n$ Latin square on $\{1, 2, \ldots, n\}$ if and only if each number in $\{1, 2, \ldots, n\}$ occurs at least $2k-n$ times in the $k \times k$ Latin square and if one further condition (relating to the diagonal elements) holds. $\square$

The reader is referred to [97] for further examples of the use of edge-colourings in problems relating to Latin squares.

Tensor Calculus

In this section we shall briefly describe R. Penrose's work on abstract tensor systems, and show how it relates to edge-colourings of planar graphs. For further details, the reader is referred to Penrose's paper [159].

We consider a simple Cartesian tensor system generated by the complex numbers as scalars, an ordinary three-dimensional Kronecker delta δ_{ab}, and the skew-symmetric (Levi-Cività) symbol ε_{abc}, where

$$
\begin{aligned}
\delta_{11} &= \delta_{22} = \delta_{33} = 1, \\
\varepsilon_{123} &= \varepsilon_{231} = \varepsilon_{312} = 1, \\
\varepsilon_{132} &= \varepsilon_{213} = \varepsilon_{321} = -1,
\end{aligned}
$$

and

all other components being zero. We can depict these by means of the following diagrams

where the factor i $(=\sqrt{-1})$ is included in order to simplify the reduction formulas obtained.

By repeatedly using well-known formulas of tensor calculus, the following reductions can be obtained:

(Intuitively, these will mean that when we come to count the total number of edge-colourings of a graph, a triangle can be removed without changing the number of edge-colourings, and a triple edge can be edge-coloured with three colours in just six ways.)

If we are given a cubic planar graph G, we can associate with it a contracted product of ε_{abc}'s, with one $i\varepsilon_{abc}$ corresponding to each vertex of G, and with a contraction occuring for each edge of G. The result of this is some complex number (actually an integer) which we denote by $K(G)$.

With this notation, Penrose proved the following result:

Theorem 9.5.   If G is a cubic planar graph, then K(G) is the total
number of ways of colouring the edges of G with three colours.

Sketch of Proof.   The tensor expression is a sum of terms, each of which
arises when one of the numbers (colours) 1, 2, 3 is assigned to each edge.
We get a factor i for each vertex where the numbers 1, 2, 3 occur in
cyclic order, −1 for each vertex where they appear in anticyclic order, and
zero if any number occurs twice.   It can be shown that each non-zero term
(that is, each proper edge-colouring) contributes exactly 1 to the sum
(this is where planarity is needed), so that the sum is precisely the number
of edge-colourings of the graph.   ☐

   We conclude this section with a simple example, to show how some of these
ideas work in practice.

Example.   Find the number of 3-colourings of the edges of the graph shown:

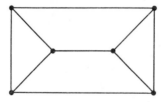

Solution.   For each edge-colouring, the contributions at the six vertices
are i (three times) and −i (three times), giving a total contribution of
$i^3 . (-i)^3 = 1$, as expected.

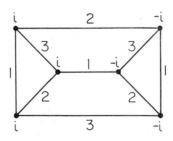

If we now apply the first of the reductions (*) to each of the two
triangles we get

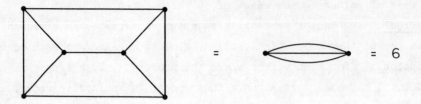

Part III—Critical graphs

In Part II we saw how graphs can be classified into two classes - those of class one, whose chromatic index equals the maximum valency, and those of class two, whose chromatic index exceeds the maximum valency by one. In this part of the book we shall restrict our attention to critical graphs, which may be described informally as graphs which are of class two, but only just! As we shall see, critical graphs have rather more structure than arbitrary graphs of class two, and we can use this fact to good advantage when proving results relating to the classification problem.

We start in Chapters 10 and 11 by defining critical graphs and deriving some of their basic properties. In particular, we shall prove an important result of Vizing (his 'Adjacency Lemma'), which will be needed several times later on. This will lead to the so-called 'Critical Graph Conjecture', which asserts that every critical graph has an odd number of vertices, a tantalizing conjecture which has so far defied all attempts at proof. In Chapter 12, we describe various constructions which can be used to generate critical graphs, and then, in Chapter 13, we obtain various results relating to the number of edges in a critical graph. These latter results are used in Chapters 14 and 15, where we exhibit a large number of critical graphs with at most ten vertices, and thereby obtain some useful evidence in support of the critical graph conjecture.

10 Critical graphs

In the study of vertex-colourings of graphs, those graphs which are critical
in some sense have played an important role. The reason for this is not
difficult to find - since every graph contains a critical graph, and since
critical graphs generally contain more structure than arbitrary graphs, we
certainly lose nothing, and often gain quite a lot, by restricting our
attention to critical graphs. The aim of this and the following chapters is
to show that a corresponding situation holds for edge-colourings.

The Definition of a Critical Graph

We define a graph G to be _critical_ (or _edge-critical_, if there is any
possibility of confusion) if G is connected and of class two, and if the
removal of any _edge_ of G lowers the chromatic index; if G has maximum
valency ρ, we shall say that G is _ρ-critical_. For example, every odd
circuit graph C_{2k+1} is a 2-critical graph, and the graph obtained by
removing any edge from the complete graph K_5 is 4-critical. On the other
hand, K_5 itself is clearly not critical, since if we remove any edge, the
remaining graph still has chromatic index 5. Tables of critical graphs
with less than ten vertices will be found in Chapters 11 and 14.

The definition of criticality can also be extended to multigraphs. If M
is a connected multigraph whose chromatic index exceeds its maximum valency,
then M is _critical_ if the removal of any edge of M lowers the chromatic
index. A table of 3-critical multigraphs with less than eight vertices will
be found in Chapter 14.

An alternative type of critical graph, which is sometimes useful, has been
introduced by Beineke and Wilson [15]. A graph G is said to be _vertex-
critical_ if G is connected and of class two, and if the removal of any
vertex of G lowers the chromatic index. (There is a corresponding
definition for multigraphs.) It is clear that every critical graph is
necessarily vertex-critical, but that the converse result is not, in general,
true; for example, K_5 is not critical, as we noted above, but it is

certainly vertex-critical. There have also been several other definitions of critical graphs, all of which are closely related to the definitions given above; the reader is referred to Hilton [110] for further details.

Critical graphs have received much more attention in the literature than vertex-critical graphs (see, for example, [14,77,128,194,195,204]); this is not really very surprising for two reasons:

(i)   critical graphs are in general much easier to deal with than vertex-critical graphs;

(ii) it is easy to show (see Exercise 10b) that a graph is critical if and only if its line-graph is vertex-critical(v) (see Chapter 3) and we can therefore use properties of vertex-critical(v) graphs to deduce properties of critical graphs.

Two Simple Results

To show how method (ii) can be used in practice, we shall consider two simple results of Vizing [194]; for each result we shall give two proofs - a direct proof, and a proof involving the line-graph which uses Theorem 3.2 on vertex-critical(v) graphs.

Theorem 10.1. If G is a ρ-critical graph, and if v and w are adjacent vertices of G, then $\rho(v) + \rho(w) \geq \rho + 2$.

First proof. Since G is ρ-critical, we can colour the edges of G-vw with ρ colours. But $\chi'(G) = \rho+1$, and so there can be no colour missing at both v and w (since otherwise that colour could be used to colour the edge vw). It follows that each of the ρ colours is needed for colouring the edges incident to v and w, so that $(\rho(v) - 1) + (\rho(w) - 1) \geq \rho$, as required. ☐

Second proof. In Theorem 3.2(b) we proved that if H is a graph which is vertex-critical(v), then the valency of each vertex of H is at least ρ. If H is taken to be the line-graph of G, and if v and w are adjacent vertices of G, then $\rho(v) + \rho(w) - 2 \geq \rho$ by a well-known property of line-graphs. ☐

Theorem 10.2. A critical graph contains no cut-vertices.

First Proof. Let G be a ρ-critical graph containing a cut-vertex v, and let $H_1, H_2, \ldots, H_k$ be the connected subgraphs of G which are obtained by removing the vertex v. Since G is critical, the subgraphs obtained by joining v to H_i, for each i, can all be edge-coloured with ρ colours; moreover, the colouring can be effected in such a way that the colours of the edges incident to v are all different. But this gives a ρ-colouring for the whole of G, contradicting the fact that G is of class two. $\square$

Second Proof. In Theorem 3.2(c) we proved that if H is a graph which is vertex-critical(v), then H cannot be expressed as the union of two graphs which intersect in a complete graph. If H is taken to be the line-graph of a critical graph G, then it follows that G cannot be expressed as the union of two graphs which intersect in a star graph $K_{1,r}$ (since $L(K_{1,r}) = K_r$). Since this is true for all values of r, G cannot contain a cut-vertex. $\square$

A Result of Vizing

In the introduction to this chapter, we asserted that every graph contains a subgraph which is critical. Our next result, which first appeared in Vizing's 1965 paper [195], asserts that every graph of class two contains a whole range of critical subgraphs, one for each possible value of ρ.

Theorem 10.3. If G is a graph of class two with maximum valency ρ, then G contains a k-critical subgraph for each k satisfying $2 \le k \le \rho$.

Proof. The result is clear for $k = \rho$, since we can obtain the required critical graph by successively removing all those edges whose deletion does not lower the chromatic index; the resulting graph clearly has the same maximum valency.

If $k < \rho$, we can take the ρ-critical graph just obtained, and colour all but one of its edges with ρ colours. If $e = vw$ is the edge which remains uncoloured, then there must be some colour α which is missing at v but not at w, and some colour β which is missing at w but not at v. If we now choose any $\rho-k$ colours different from α or β, and if we remove all those edges of G which are coloured with any of these $\rho-k$ colours, we obtain a subgraph which has maximum valency k and chromatic index $k+1$. By removing edges, as in the first part of this proof, we obtain the required k-critical subgraph. $\square$

The definition of a critical graph, given at the beginning of this chapter, states that if any single edge is deleted, then the chromatic index of the

68

graph is reduced by one. Our next result shows that we can say rather more than this; its proof is immediate and will be left to the reader (see Exercise 10b).

Theorem 10.4. If G is a critical graph, and if I is any set of independent edges of G, then
(i)   there exists a $(\rho+1)$-colouring of the edges of G in which I is a colour class;
(ii) $\chi'(G-I) = \chi'(G) - 1$. $\square$

Note, as a special case of Theorem 10.4, that the removal of a 1-factor from a ρ-critical graph (with $\rho \geq 3$) leaves a graph which is still of class two.

The Non-Regularity of Critical Graphs

We conclude this chapter by proving that the only critical graphs which are regular are the odd circuit graphs C_{2k+1}.

Theorem 10.5. There are no regular ρ-critical graphs with $\rho \geq 3$.
Proof. By Corollary 6.4 there are no regular ρ-critical graphs ($\rho \geq 3$) of odd order. If G is a regular ρ-critical graph of even order, and if H is the graph obtained from G by inserting a new vertex z into the edge $e = vw$ of G, then $\chi'(H) = \rho+1$, by Corollary 6.5. Now $\chi'(G-e) = \rho$, since G is critical, and in any colouring of the edges of G-e there is at least one colour α missing at v and at least one colour β missing at w; we may assume that $\alpha \neq \beta$, since otherwise we could immediately obtain a ρ-colouring of the edges of G. It follows that we can obtain a ρ-colouring of the edges of H by colouring the edge vz with colour α and the edge wz with colour β. This contradiction establishes the result. $\square$

Exercises.

10a (i)   Which of the eight graphs in Figure 6.1 are critical graphs?
   (ii)   Prove directly (without using Theorem 10.5) that the Petersen graph is not critical. Is it vertex-critical?

10b (i)   Prove that a connected graph is critical if and only if its line-graph is vertex-critical(v).

(ii) Prove Theorem 10.4.

10c (i) Find an example of a 3-critical graph which is not Hamiltonian.

(ii) Find an example of a 4-valent graph of class two which contains two 4-critical subgraphs of different orders. (Fiorini [69,75].)

10d Prove that, if G is a ρ-critical graph with edge-independence number α, then G has at most $\alpha\rho+1$ edges. Construct an example to show that this bound can actually be attained.

10e Let G be a ρ-critical graph, and let $e = vw$ be an edge of G. Suppose that we are given a ρ-colouring of the edges of $G-e$, and that Θ_v and Θ_w denote the sets of colours which appear at v and w, respectively. Prove that

(i) $|\Theta_v \cup \Theta_w| = \rho;$

(ii) $|\Theta_v \cap \Theta_w| = \rho(v) + \rho(w) - \rho - 2;$

(iii) $|\Theta_v \setminus \Theta_w| = \rho + 1 - \rho(w);$

(iv) $|\Theta_w \setminus \Theta_v| = \rho + 1 - \rho(v).$ (Berge [17].)

10f (i) A connected graph G is said to be <u>index-critical</u> if, for every proper subgraph H of G, $\chi'(H) < \chi'(G)$. Prove that if G is index-critical then G is critical, and conversely.

(ii) A connected graph G is said to be <u>class-critical</u> if G is of class two but, for every proper subgraph H of G, H is of class one. Prove that if G is class-critical, then G must be an odd circuit. (Hilton [110].)

10g (i) Let G be the graph obtained by inserting a vertex v into any edge of the complete bipartite graph $K_{k,k}$. Show that, if e is any edge of G, then there is a Hamiltonian circuit in G which includes both the vertex v and the edge e, and deduce that G is k-critical.

(ii) Let G be the graph obtained by inserting a vertex into any edge of the complete graph K_{2k}. Use the method of part (i) to prove that G is $(2k-1)$-critical. Show further that if $\tilde{G}$ is any graph obtained from K_{2k} by first removing a 1-factor and then inserting a vertex into any of the remaining edges, then $\tilde{G}$ is also a critical graph. (Fiorini [68].)

11 The structure of critical graphs

In the previous chapter we defined a graph to be critical if it is connected and of class two, and if the removal of any edge results in a graph of lower chromatic index. At first sight, it seems that the lowering of the chromatic index can occur in either of two different ways:

(i)   the removal of an edge produces a graph of class one with the same maximum valency;

(ii) the removal of an edge produces a graph which is still of class two, but with smaller maximum valency.

It is clear that the second case could arise only if our original graph had at most two vertices of maximum valency, and it is one of the objects of this chapter to show that this can never happen. We shall deduce this result as a special case of an extremely important result of Vizing, which we have called his 'Adjacency Lemma' in order to distinguish it from his Classification Theorem in Chapter 5. This lemma will be used many times throughout the rest of this book, and in the second half of this chapter we illustrate its use by deriving several properties of 3-critical graphs.

Vizing's Adjacency Lemma

We start by stating a result which shows that case (ii) above cannot arise. It is due to Vizing, and appeared in [194,195]; we omit its proof since it is a special case of Theorem 11.2 which we shall prove in full.

Theorem 11.1. If G is a critical graph, then every vertex of G is adjacent to at least two vertices of maximum valency; in particular, G contains at least three vertices of maximum valency. □

In Vizing's 1965 paper [195], he generalized this result and proved his 'Adjacency Lemma' (Theorem 11.2), which clearly includes Theorem 11.1 as a special case. Before proving it we shall need to introduce the idea of a 'fan-sequence'. Let w be a vertex of a graph G whose edges have been coloured in such a way that adjacent edges are assigned distinct colours.

A <u>fan-sequence</u> F at w with initial edge wa_1 is a sequence of distinct edges of the form $wa_1, wa_2, wa_3, \ldots$, such that for each $i \geq 1$, the edge wa_{i+1} is coloured with a colour missing from the vertex a_i.

<u>Theorem 11.2</u> (Vizing's Adjacency Lemma). Let G be a ρ-critical graph, and let v and w be adjacent vertices of G with $\rho(v) = k$. Then

(*i*) if $k < \rho$, then w is adjacent to at least $\rho-k+1$ vertices of valency ρ;

(*ii*) if $k = \rho$, then w is adjacent to at least two vertices of valency ρ.

<u>Proof.</u> Since G is critical, we can find a ρ-colouring of the edges of $G-e$, where $e = vw$. We may assume that each of the $\rho-k+1$ colours missing from the vertex v appears at the vertex w, since otherwise we could immediately obtain a ρ-colouring of the edges of G; it follows that $\rho(w) \geq \rho-k+2$.

We now make two assertions:

(1) If F and F' are two distinct fan-sequences at w in $G-e$, and if the initial edges of F and F' are distinct and coloured with colours which appear at w but not at v, then F and F' have no edges in common.

(2) If $F = \{wa_1, \ldots, wa_s\}$ is a fan-sequence of maximal length at w, starting with an edge wa_1, whose colour does not appear at v, then the vertex a_s has valency ρ.

To prove (1), let $F = \{wa_1, \ldots, wa_s\}$ and $F' = \{wb_1, \ldots, wb_t\}$ be two fan-sequences at w, and suppose that $wa_s = wb_t$. Without loss of generality, we may assume that all of the other edges are distinct, and that subject to these conditions, F and F' have been chosen to be of minimal size. Since either F or F' must contain at least two edges, we may suppose that $s \geq 2$. Let α_i ($i = 1, 2, \ldots, s$) and β_i ($i = 1, 2, \ldots, t$) be the colours of the edges in F and F', respectively; then $\beta_t = \alpha_s$, and if $t \geq 2$ then α_s does not appear at either a_{s-1} or b_{t-1} (see Figure 11.1). Since F and F' are of minimal size, the only colours which appear at w but not at v are α_1 and β_1. So let γ be a colour which appears at v but not at w, and let Γ be a chain of maximum length consisting of edges alternately coloured α_s and γ and with initial edge wa_s ($= wb_t$). There are now two cases to consider:

Case (i): <u>if $t \geq 2$</u>, then Γ may terminate at either a_{s-1} or b_{t-1}, or at

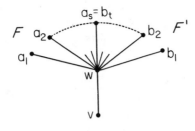

Fig. 11.1

neither of these. In the first case, we can interchange the colours of the
edges in Γ, and recolour each edge wb_i with colour β_{i+1}, for
$i = 1,\ldots,t-1$; we can then colour the edge vw with colour β_1. In the
second and third cases, we can interchange the colours of the edges in Γ,
and recolour each edge wa_i with colour α_{i+1}, for $i = 1,\ldots,s-1$; we can
then colour the edge vw with colour α_1.

Case (ii): <u>if $t = 1$</u>, then Γ may terminate either at v or at some other
vertex. In the first case, we can interchange the colours of the edges in
Γ, and recolour each edge wa_i with colour α_{i+1}, for $i = 1,\ldots,s-1$; we
can then colour the edge vw with colour α_1. In the second case, we can
interchange the colours of the edges in Γ, and we can then colour the edge
vw with colour β_1. This concludes the proof of (1).

To prove (2), we shall prove the stronger result that if α_i is the colour of wa_i, for each i, then F contains *no* edge wa_k with the property
that, for some $i < k$, the colour α_i is missing from the vertex a_k. So
suppose that for some $i < k$ the colour α_i is missing from the vertex a_k,
and let γ be a colour which appears at v but not at w.

<u>If $i \geq 2$</u>, then we may assume the existence of two chains with edges
alternately coloured α_i and γ, as follows:

(i)   we may assume the existence of an (α_i,γ)-chain starting from the vertex
a_{i-1} with an edge coloured γ, and ending with the edge a_iw, coloured α_i
(see Figure 11.2); for otherwise we could recolour wa_{i-1} with colour γ,
interchange the colours on the maximal (α_i,γ)-chain starting at a_{i-1},
recolour the edge wa_i with colour α_{i+1}, for $i = 1,\ldots,i-2$, and finally
colour the edge vw with colour α_1.

(ii)   we may assume the existence of an (α_i,γ)-chain starting from the vertex

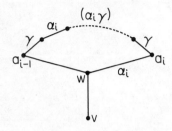

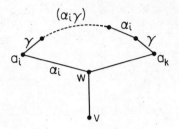

Fig. 11.2   Fig. 11.3

a_k with an edge coloured γ, and ending either at the vertex w with the edge a_iw, coloured α_i (see Figure 11.3), or at the vertex a_{i-1}, ending with an edge coloured γ; for otherwise, we could recolour wa_k with colour γ, interchange the colours on the maximal (α_i,γ)-chain starting at a_k, recolour the edge wa_i with colour α_{i+1}, for $i = 1,\ldots,k-1$, and finally colour the edge vw with colour α_1.

However, the simultaneous existence of these two (α_i,γ)-chains is clearly impossible, and this gives the required contradiction.

If $i = 2$, we can recolour wa_i, for $i = 1,\ldots,k-1$, colour wa_k with colour γ, interchange the colours on the maximal (α_i,γ)-chain from a_k to a_1, and finally colour the edge vw with colour α_1. This concludes the proof of (2).

We can now complete the proof of the theorem. For each of the $\rho-k+1$ colours appearing at w but not at v, there is a fan-sequence of maximal size at w, starting with an edge of that colour. By (1) these fan-sequences are all disjoint, and by (2) each one ends with an edge incident to a vertex of valency ρ. The result now follows immediately. $\square$

By taking v to be a vertex of minimum valency σ, and w to be a vertex of maximum valency ρ, we immediately obtain the following corollary of Theorem 11.2:

Corollary 11.3. If G is a ρ-critical graph with minimum valency σ, then G contains at least $\rho-\sigma+2$ vertices of maximum valency. $\square$

In later chapters we shall describe how Vizing used his Adjacency Lemma to obtain bounds on the number of edges and the lengths of circuits of critical graphs. We shall also use it to derive results on the Critical Graph Conjecture, and on the chromatic index of planar graphs.

The Adjacency Lemma for Multigraphs

Since critical multigraphs have received so much attention in the literature, it is natural to ask whether there are multigraph analogues of Theorems 11.1 and 11.2, and Corollary 11.3. That such analogues exist was proved by Andersen [2,3] in the case where the chromatic index of the multigraph M is equal to its maximum enlarged valency $\rho*$ (see Theorem 5.4 and the definitions preceding it). In particular, he proved the following results:

Theorem 11.4. Let M be a critical multigraph with maximum enlarged valency $\rho*$, and suppose that $\chi'(M) = \rho*$. Then
(i)   every vertex v of M is adjacent to at least two vertices w for which $\rho(w) + \mu(vw) = \rho*$;
(ii)   if v and w are adjacent vertices of M with $\rho(v) + \mu(vw) = k$, then w is adjacent to at least $\rho*-k+1$ vertices z such that $\rho(z) + \mu(wz) = \rho*$;
(iii) if M has minimum enlarged valency $\sigma*$, then M contains at least $\rho*-\sigma*+1$ vertices of maximum enlarged valency. □

3-critical Graphs

We conclude this chapter by looking at ρ-critical graphs with a small value of ρ. There are clearly no 1-critical graphs, and the only 2-critical graphs are the odd circuits C_{2k+1}. The first case of interest is the investigation of 3-critical graphs, and, as the reader will probably have gathered from Chapter 6, even this case is way beyond the power of present-day techniques. However, we can still use Theorems 11.1 and 11.2 to obtain some useful properties of 3-critical graphs, and we shall show how I.T. Jakobsen [128] was thereby able to obtain all of the 3-critical graphs with less than ten vertices.

Jakobsen started by observing that since a critical graph can contain no cut-vertex, every vertex of a 3-critical graph G must have valency 2 or 3.

If n_2 and n_3 denote the number of vertices of valency 2 and 3, respectively, and if n is the total number of vertices, then clearly $n = n_2 + n_3$. But by Theorem 11.2, every vertex of valency 2 is adjacent to two vertices of valency 3, and every vertex of valency 3 must be adjacent to at most one vertex of valency 2. It follows that $n_3 \geq 2n_2$, and hence that $n_2 \leq \frac{1}{3}n$ and $n_3 \geq \frac{2}{3}n$. The number of edges of G is therefore equal to $\frac{1}{2}(2n_2 + 3n_3) = (n - n_3) + \frac{3}{2}n_3 \geq \frac{4}{3}n$.

In order to obtain an upper bound for the number of edges of G, we observe simply that G cannot be regular, by Theorem 10.5, and so must contain at least one vertex of valency 2. It follows that the number of edges of G is at most $\frac{1}{2}(2 + 3(n-1)) = \frac{1}{2}(3n - 1)$. We summarize these results as a theorem:

Theorem 11.5. If G is a 3-critical graph with n vertices and m edges, then $\frac{4}{3}n \leq m \leq \frac{1}{2}(3n - 1)$. □

In the same paper, Jakobsen proved that no 3-critical graph G can contain exactly two vertices of valency 2. But if G is of even order, then the number of vertices of valency 3 must be even, and so the number of vertices of valency 2 must also be even. It follows that G has at least four vertices of valency 2, so that $4 \leq \frac{1}{3}n$ – that is, $n \geq 12$. We therefore have the following result:

Theorem 11.6. There are no 3-critical graphs of order 4, 6, 8 or 10. □

This intriguing fact led Jakobsen [127] to formulate the following conjecture, also conjectured independently (but for vertex-critical graphs) by Beineke and Wilson [15]:

The Critical Graph Conjecture. Every critical graph has an odd number of vertices.

Jakobsen devoted the rest of his paper to a search for all 3-critical graphs with 5, 7 and 9 vertices, and we present his results in Figure 11.4. We shall not give any of his arguments here, since we shall be returning to such questions in Chapters 14 and 15, where we prove the non-existence of

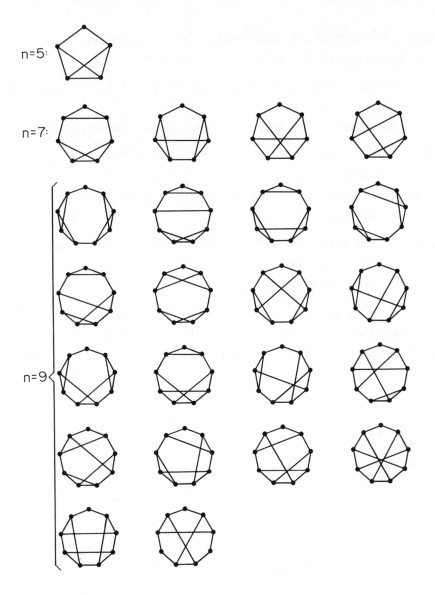

n=5:

n=7:

n=9

Fig. 11.4

ρ-critical graphs with 2, 4, 6, 8 or 10 vertices (for any value of ρ) and 3-critical graphs with 12 vertices.

Exercises.

11a Verify the statements of Theorem 11.2 and Corollary 11.3 for the graphs
     (c) and (d) of Figure 6.1.

11b   Show that the Critical Graph Conjecture is equivalent to the correspond-
      ing conjecture for vertex-critical graphs.

11c   (i)   Verify the bounds in Theorem 11.5 for each of the graphs in
             Figure 11.4.
      (ii)   Give examples to show that these bounds can actually be attained.

11d   (i)   For each odd number n ≥ 5, construct a 3-critical graph with n
             vertices.
      (ii)   Prove that a 3-critical graph cannot contain exactly two vertices
             of valency 2.   (Jakobsen [128].)

11e   Let G be a graph, and let H be the subgraph of G induced by the
      vertices of maximum valency. Prove that if H contains no circuits,
      then G is of class one.

11f   Let M be a ring-multigraph (see Exercise 5i) with m edges and
      maximum valency ρ. Prove that if M is critical, and if
      $\chi'(M) > (m\rho + m - 3)/(m - 1)$, then M has at most m - 2 vertices.
                                                                (Jakobsen [129].)

12 Constructions for critical graphs

In Chapter 10 we showed that if G is any graph of class two with maximum valency ρ. then G must contain a k-critical subgraph for each k satisfying $2 \leq k \leq \rho$. Later, in Exercise 10g, we saw how critical graphs could be obtained from certain complete graphs and complete bipartite graphs, and at the end of Chapter 11, we exhibited all of the 3-critical graphs of order $n \leq 9$. But apart from these, we have up to now produced very few critical graphs to work with. The object of this chapter is to set this right by describing various constructions which can be used to obtain new critical graphs from previously-known ones. These constructions will be of two distinct types:

Type (i): Constructions which start from ρ-critical graphs, and yield ρ'-critical graphs of the same order, where $\rho' > \rho$;

Type (ii): Constructions which start from ρ-critical graphs, and yield ρ-critical graphs of larger order.

Constructions Involving Circuit Graphs

We start by describing a type (i) construction which starts from a given set of 3-critical graphs and yields ρ-critical graphs for any desired value of ρ. The basic idea is that if we have two 3-critical graphs each of which can be obtained from the odd circuit C_{2s+1} by adding an independent set of s edges, and if the two independent sets are disjoint, then we can add *both* independent sets of edges to produce a 4-critical graph. This procedure is illustrated in Figure 12.1 in the case s = 3. This construction can be gen-

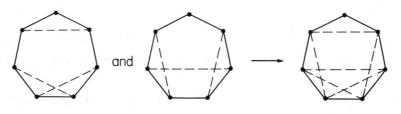

Fig. 12.1

eralized to any number of 3-critical graphs, as follows:

Theorem 12.1. For $i = 1, 2, \ldots, t$, let G_i be a 3-critical graph obtained from C_{2s+1} by adding an independent set S_i of s edges. If $t \leq 2s - 4$, and if the sets S_i are pairwise-disjoint, then the graph G obtained from C_{2s+1} by adding *all* of the sets S_i is a $(t+2)$-critical graph.

Proof. It is easy to see that G has $2s + 1$ vertices, $(t+2)s + 1$ edges, and maximum valency $t + 2$. It follows from Theorem 6.1 that G is of class two, and hence that $\chi'(G) = t + 3$. In order to prove that G is $(t+2)$-critical, we must prove that for any edge e, $\chi'(G-e) < \chi'(G)$. There are two cases to consider:

(i) If e is any edge of the boundary circuit C_{2s+1}, then we can colour the edges of $C_{2s+1}-e$ with two colours. If we now assign a separate colour to each of the sets S_i, we immediately obtain a $(t+2)$-colouring of the edges of $G-e$. This completes the proof in this case.

(ii) If e is in the set S_i, then the graph G_i-e is 3-colourable, since G_i is 3-critical. If we now assign a separate colour to each of the sets S_j $(j \neq i)$, we immediately obtain a $(t+2)$-colouring of the edges of $G-e$. This completes the proof. $\square$

Using constructions such as these, Buttiġieġ, Fiorini and Galea have developed a computer programme (see [68]) which generates ρ-critical graphs for any desired value of ρ. By examining such constructions, it is possible to prove the existence of ρ-critical graphs *containing a vertex of valency 2*, for any desired value of ρ (see Fiorini and Wilson [77]). This is in sharp contrast to the situation which holds for vertex-colourings, where every vertex-critical(v) graph with chromatic number k is necessarily $(k-1)$-connected; all we can say in general about ρ-critical graphs (in the edge-colouring sense) is that they are 2-connected. A precise statement of this connectivity result is as follows:

Theorem 12.2. If n is any odd integer $(n \geq 7)$, then there exists a ρ-critical graph of order n containing a vertex of valency 2, for ρ satisfying $2 \leq \rho \leq n - 4$.

Proof. Let $n = 2s+1$. In order to prove this result, we shall exhibit $2s-5$ 3-critical graphs of the kind described in Theorem 12.1, in each of which the vertex 2s+1 has valency 2; the result then follows from Theorem 12.1, by taking these graphs to be the graphs G_i. The graphs G_i are obtained as follows:

Type 1. We first construct $s - 4$ 3-critical graphs $X_1, \ldots, X_{s-4}$, as follows: for each k, the graph X_k is obtained by adding to C_{2s+1} those edges which join the pairs of vertices

$$1 \quad \text{and} \quad k+2; \quad s+k \quad \text{and} \quad 2s; \quad s+k+1 \quad \text{and} \quad 2s-1;$$
$$k+2-i \quad \text{and} \quad k+2+i \quad (\text{for} \quad i = 1,2,\ldots,k);$$
$$s+k-j \quad \text{and} \quad s+k+j+1 \quad (\text{for} \quad j = 1,2,\ldots,s-k-3).$$

(This procedure is illustrated in Figure 12.2 for the case $s = 6$, $k = 2$.)

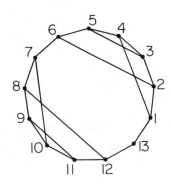

Fig. 12.2

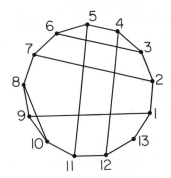

Fig. 12.3

Type 2. We next construct $s - 4$ 3-critical graphs $Y_1, \ldots, Y_{s-4}$, as follows: for each k, the graph Y_k is obtained by adding to C_{2s+1} those edges which join the pairs of vertices

$$1 \quad \text{and} \quad s+k+1; \quad k+2 \quad \text{and} \quad 2s; \quad k+3 \quad \text{and} \quad 2s-1;$$
$$k+2-i \quad \text{and} \quad k+3+i \quad (\text{for} \quad i = 1,2,\ldots,k)$$
$$s+k-j+1 \quad \text{and} \quad s+k+j+1 \quad (\text{for} \quad j = 1,2,\ldots,s-k-3).$$

(This procedure is illustrated in Figure 12.3 for the case $s = 6$, $k = 2$.)

Type 3. Finally, we construct the three 3-critical graphs Z_1, Z_2 and Z_3, obtained by adding to C_{2s+1} those edges which join the pairs of vertices

(for Z_1) $s-1$ and $2s$; s and $2s-1$;

$$i \quad \text{and} \quad 2s-i-1 \quad (i = 1,2,\ldots,s-2).$$

(for Z_2) s and $2s$; i and $2s-i$ $(i = 1,2,\ldots,s-1)$

(for Z_3)   1 and s+1; i+1 and 2s−i+1   (i = 1,2,...,s−1).
(These graphs are illustrated in Figure 12.4 for the case s = 6.)   □

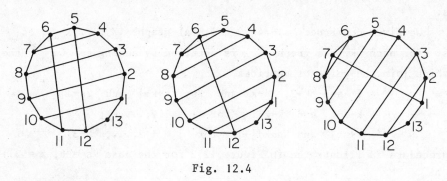

Fig. 12.4

The Hajós-Union Construction

We turn now to a type (ii) construction – that is a construction which
produces ρ-critical graphs from ρ-critical graphs of smaller order. One of
the most useful of these constructions is due to Jakobsen [127], and involves
the Hajós-union of two graphs (defined in Chapter 2).

Theorem 12.3. Let G   and   G' be two ρ-critical graphs, and let K be any
Hajós-union of G   and   G' obtained by identifying two vertices, the sum of
whose valencies does not exceed ρ+2.   Then K is also ρ-critical.
Proof. We show first that K is of class two. Suppose that the identified
vertices of G and G' are v and v' respectively, and that K is obtained by removing vz and v'z', and adding the edge zz'. Since the sum
of the valencies of v and v' (in G and G') does not exceed ρ + 2,
the maximum valency of K must be ρ. If K is ρ-colourable, and if the
edge zz' is assigned the colour α, then some edge of G incident to v
must also have colour α, since otherwise we could colour the edges of G
with just ρ colours. Similarly, some edge of G' incident to v' must
have colour α, and so (since v and v' have been identified) there are
two adjacent edges coloured α. This contradiction establishes that K is
of class two.

In order to show that K is ρ-critical, we must prove that K−e is ρ-
colourable for each edge e. If e = zz', this is clear, since the ρ-colour-
ings of G−vz and G'−v'z' can be combined to give a ρ-colouring of the
edges of K−zz'. So we may assume that e is an edge of G−vz (since the

82

corresponding argument for $G'-v'z'$ is exactly the same). Now the graph $(G'-v'z') \cup zz'$ is ρ-colourable, and in each ρ-colouring, the colour α of the edge zz' appears at the vertex v'. But $G-e$ is also ρ-colourable, and so any ρ-colouring of $G-e$ gives rise to a ρ-colouring of $(G-vz-e) \cup zz'$, in which the colour α of zz' is missing from v. It follows that the ρ-colourings of $(G-vz-e) \cup zz'$ and $(G'-v'z') \cup zz'$ can be combined to give a ρ-colouring of $K-e$. This completes the proof. $\square$

Using Theorem 12.3, it is a straightforward matter to characterize critical graphs or multigraphs K which are 'separable by two independent edges' - this means that it is possible to disconnect K by removing two non-adjacent edges, but not by removing just one edge. This result will be used in Chapter 15, and it is due to Jakobsen:

Theorem 12.4. A ρ-critical multigraph K is separable by two independent edges if and only if K can be expressed as a Hajós-union of two ρ-critical multigraphs G and G' by identifying a vertex of valency 2 in G with some vertex in G'. $\square$

Exercises.

12a Check that the graphs X_i, Y_i and Z_i in the proof of Theorem 12.2 are all 3-critical, and that the corresponding sets S_i are pairwise-disjoint.

12b Prove that if G is any graph obtained from C_{2s+1} by adding t disjoint independent sets of s edges (where $t \leq 2s-2$), then G is vertex-critical. (Beineke and Wilson [15].)

12c Let G be a ρ-critical graph of order $2s+1$, and suppose that (i) G does not contain three independent edges, and (ii) if m is the number of edges of G, then $\{\frac{m}{s}\} = \rho+1$. Prove that if G' is any graph obtained from G by adding a set of s independent edges, then G' is $(\rho+1)$-critical. (Jakobsen [129].)

12d Let G be a ρ-critical graph (where ρ is odd), let $H = K_{\rho+1}$, and let K be the graph obtained from G and H by the following procedure:
(i) choose a vertex v of valency ρ in G, and label its neighbours $v_1, v_2, \dots, v_\rho$ in an arbitrary manner;

(ii) choose a vertex w in H, and label its neighbours w_1, w_2,...,w_ρ in an arbitrary manner;

(iii) delete the vertex v from G and the vertex w from H, and join each v_i to the corresponding w_i.

Using the result of Exercise 6h, prove that K is ρ-critical, and explain why the result is false if ρ is even.   (Fiorini [68].)

12e Prove that the result of Exercise 12d is unchanged if we replace '$H = K_{\rho+1}$' by '$H = K_{\rho,\rho}$'. What other graphs can be used for the graph H?

12f Prove the following 'converse' of the result of Exercise 12d: Let ρ be odd, and let K be a ρ-critical graph which is separable by a set E of ρ independent edges into two graphs G and H. If the valency in K of each vertex in H is ρ, prove that the graph obtained from K by contracting H to a single vertex is also ρ-critical.

13 Bounds on the number of edges

In Chapter 11 we saw how I.T. Jakobsen used Vizing's Adjacency Lemma to obtain a lower bound for the number of edges of a 3-critical graph. Our aim in this chapter is to describe a variety of methods which can be used to obtain both upper and lower bounds for the number of edges of an arbitrary critical graph. Several of the results we obtain here will be used in the following two chapters, where we discuss critical graphs of small order.

An Upper Bound

In this section we present an upper bound for the number of edges of a critical graph which is obtained by using Vizing's Adjacency Lemma.

Theorem 13.1. If G is a critical graph with n vertices, m edges, maximum valency ρ and minimum valency σ, then

$$m \leq \tfrac{1}{2}(n - 1)\rho + 1, \quad \text{if } n \text{ is odd,}$$

and

$$m \leq \tfrac{1}{2}(n - 2)\rho + \sigma - 1, \text{ if } n \text{ is even.}$$

Proof. If n is odd, then the inequality $m \leq \tfrac{1}{2}(n - 1)\rho + 1$ follows immediately from Theorem 6.1.

If n is even, let v be a vertex of valency σ, and let v_1 be a vertex of valency ρ adjacent to v; such a vertex v_1 exists by Theorem 11.1. Since G is critical, the graph obtained from G by deleting the edge vv_1 is ρ-colourable, and in any such colouring there is a colour missing at v_1. We may assume that this missing colour is used for some edge vv_2, since otherwise we could use it to colour the edge vv_1, contradicting the fact that G is of class two. It follows that the multigraph G' obtained from G by deleting v and adding an edge v_1v_2 is ρ-colourable. If v_1 and v_2 are joined by two edges coloured α and β, say, we can transform G' into a graph by replacing the double-edge by a copy of $K_{\rho,\rho}$ with two non-adjacent edges removed, as in Figure 13.1. In

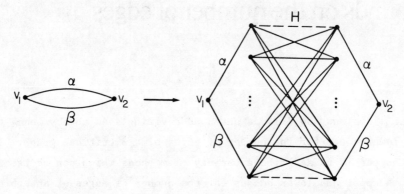

Fig. 13.1

any case, we end up with a graph $\tilde{G}$ which is of class one and has odd order. It follows that, if $\tau(G)$ and $\tau(\tilde{G})$ are the total deficiencies of G and $\tilde{G}$, then

$$\tau(\tilde{G}) = \tau(G) - (\rho - \sigma) + (\sigma - 2),$$

and

$$\tau(\tilde{G}) \geq \rho, \quad \text{by Corollary 6.2.}$$

So

$$\tau(G) \geq 2(\rho - \sigma + 1),$$

that is,

$$m \leq \tfrac{1}{2}(n - 2)\rho + \sigma - 1, \quad \text{as required.} \quad \square$$

Note that we can use the result of Theorem 13.1 to give an alternative proof of Theorem 10.5 (that if $\rho \geq 3$ then G cannot be regular), since for every regular ρ-valent graph we must have $m = \tfrac{1}{2}n\rho$.

Lower Bounds

We now turn our attention to lower bounds for the number of edges of a ρ-critical graph. The first such bound was obtained by Vizing [195] as a simple consequence of his Adjacency Lemma.

Theorem 13.2. If G is a ρ-critical graph with m edges, then

$$m \geq \tfrac{1}{8}(3\rho^2 + 6\rho - 1).$$

Proof. If σ is the minimum valency in G, then G must contain at least $\rho - \sigma + 2$ vertices of valency ρ, by Corollary 11.3. But G must have at least $\rho + 1$ vertices, and so the number of edges of G must satisfy

$$2m \geq \rho(\rho - \sigma + 2) + \sigma((\rho+1) - (\rho-\sigma+2))$$
$$= \rho(\rho - \sigma + 2) + \sigma(\sigma - 1).$$

This expression is smallest when $\sigma = [\frac{1}{2}(\rho + 1)]$, as may easily be seen by taking derivatives. It follows that

$$2m \geq \rho(\frac{1}{2}\rho + \frac{3}{2}) + \frac{1}{4}(\rho + 1)(\rho - 1)$$
$$= \frac{1}{4}(3\rho^2 + 6\rho - 1),$$

as required. ☐

This bound is a good one when the number of vertices is small. For example, if $G = K_5 - e$, then Theorem 13.2 gives 9 (the correct answer) as the lower bound. On the other hand, if the order of G is large compared with ρ, then this bound can be very poor. It is therefore natural to ask whether there are lower bounds for m which involve both ρ and n. One simple bound of this kind is given by our next theorem:

Theorem 13.3. If G is a ρ-critical graph of order n, then G has at least $2n(\rho - 1)/\rho$ edges.

Proof. If G has n_ρ vertices of valency ρ, then

$$2m \geq \rho n_\rho + 2(n - n_\rho),$$

since the valency of each vertex is at least two. But by Corollary 11.1, each vertex of G is adjacent to at least two vertices of valency ρ, and so $\rho n_\rho \geq 2n$. The required result follows by combining these two inequalities. ☐

If we take $\rho = 3$ in the statement of Theorem 13.3, we see that every 3-critical graph has at least $\frac{4}{3}n$ edges - this is Jakobsen's lower bound which we have already obtained in Chapter 11. However, for $\rho \geq 3$, we can improve on the bound given by Theorem 13.3. Since the argument is somewhat complicated, we shall give the full details only for the case $\rho = 4$.

Theorem 13.4. If G is a 4-critical graph of order n, then G has at least $\frac{5}{3}n$ edges.

Proof. If n_j is the number of vertices of valency j, for $j = 2, 3, 4$, then

$$n = n_2 + n_3 + n_4, \quad \text{and} \quad 2m = 2n_2 + 3n_3 + 4n_4,$$

so that

$$2n_2 + n_3 = 4n - 2m.$$

Also, since the valency of each vertex is at least two, we have $2m \geq 2n + 2n_4$, so that $n_4 \leq m - n$.

But, by Theorem 11.1, every vertex of valency two is adjacent to two vertices of valency 4, and by Vizing's Adjacency Lemma, each of these is adjacent to three further vertices of valency 4. It follows that, if $n_4(r,s)$ is the number of vertices of valency 4, adjacent to exactly r vertices of valency 2, and exactly s vertices of valency 3, then $2n_2 = n_4(1,0)$.

Similarly, since each vertex of valency 3 is adjacent to at least two vertices of valency 4, and since each of these is adjacent to two further vertices of valency 4, we have $2n_3 \leq n_4(1,0) + 2n_4(0,2)$. It follows that

$$2n_2 + n_3 \leq n_4(1,0) + \tfrac{1}{2}n_4(0,1) + n_4(0,2) \leq n_4,$$

so that

$$4n - 2m \leq m - n,$$

giving the required result. □

In order to generalize this method to larger values of ρ, we need to find an analogue of the above inequality $2n_2 + n_3 \leq n_4$. This analogue is provided by the following useful result, which will be needed over and over again in the following two chapters:

Theorem 13.5. Let G be a ρ-critical graph, and let n_j be the number of vertices of valency j, for $j = 2,3,\ldots,\rho$. Then for each k satisfying $2 \leq k \leq \rho-1$, we have

$$\sum_{j=2}^{k} \frac{n_j}{j-1} \leq \tfrac{1}{2}n_\rho.$$

Proof. To each vertex v of valency ρ in G, assign a $(k-1)$-tuple

$(i_2, i_3, \ldots, i_k)$, where for each value of t, i_t is the number of vertices of valency t adjacent to v, and let $n_\rho(i_2, i_3, \ldots, i_k)$ be the number of vertices of valency ρ associated with this $(k-1)$-tuple. Then since each vertex of G is adjacent to at least two vertices of valency ρ (by Theorem 11.1), we have

$$2n_j \le \sum_{(k)} i_j \cdot n_\rho(i_2, i_3, \ldots, i_k),$$

where the summation extends over all $(k-1)$-tuples associated with any vertex of valency ρ.

It follows that

$$\sum_{j=2}^{k} \frac{2n_j}{j-1} \le \sum_{j=2}^{k} \sum_{(k)} \frac{i_j}{j-1} \cdot n_\rho(i_2, \ldots, i_k)$$

$$= \sum_{(k)} n_\rho(i_2, \ldots, i_k) \sum_{j=2}^{k} \frac{i_j}{j-1}$$

$$\le \sum_{(k)} n_\rho(i_2, \ldots, i_k) \sum_{j=2}^{k} \frac{i_j}{q-1},$$

where q is the smallest index of all *non-zero* elements of the $(k-1)$-tuple $(i_2, \ldots, i_k)$. But the vertex v is adjacent to at least $\rho - q + 1$ vertices of valency ρ, by Vizing's Adjacency Lemma, and so must be adjacent to at most $\rho - (\rho - q + 1) = q - 1$ vertices whose valency is less than ρ. It follows that

$$i_2 + i_3 + \ldots + i_k \le q - 1,$$

and hence that

$$\sum_{j=2}^{k} \frac{2n_j}{j-1} \le \sum_{(k)} n_\rho(i_2, \ldots, i_k) \le n_\rho, \quad \text{as required.} \quad \square$$

Using the result of Theorem 13.5, together with the method of Theorem 13.4, Fiorini [71] was able to obtain the following lower bound for the number of edges of a ρ-critical graph:

<u>Theorem 13.6</u>. If G is a ρ-critical graph with n vertices and m edges, then

$$m \geq \tfrac{1}{4}n(\rho + 1), \text{ if } \rho \text{ is odd,}$$

and

$$m \geq \tfrac{1}{4}n(\rho + 2), \text{ if } \rho \text{ is even.} \quad \square$$

It is likely that the bounds given in Theorem 13.6 can be improved considerably. In fact, Vizing [196] has formulated the following conjecture, which still remains unproved:

<u>Vizing's Conjecture</u>. Every ρ-critical graph of order n has at least $\tfrac{1}{2}(n\rho - n + 3)$ edges.

It can easily be shown (see Exercise 13d) that if Vizing's Conjecture is correct, then every *planar* graph with $\rho \geq 7$ is necessarily of class one; this extends the result of Theorem 6.6 which states the analogous result for $\rho \geq 8$. We shall give a proof of Theorem 6.6 when we discuss planar graphs in Chapter 16.

We conclude this chapter by summarizing the various lower bounds discussed in this chapter; where appropriate, these bounds are rounded up to the next integer:

	Theorem 13.2 $\tfrac{1}{8}(3\rho^2+6\rho-1)$	Theorem 13.3 $\tfrac{2n}{\rho}(\rho-1)$	Theorems 11.5 13.4 etc. (Small values of ρ)	Theorem 13.6 $\tfrac{1}{4}n(\rho+i)$	Vizing's Conjecture $\tfrac{1}{2}(n\rho-n+3)$
$\rho=2$	3	n	n	n	$\tfrac{1}{2}(n+3)$
$\rho=3$	6	$\tfrac{4}{3}n$	$\tfrac{4}{3}n$	n	$n+2$
$\rho=4$	9	$\tfrac{3}{2}n$	$\tfrac{5}{3}n$	$\tfrac{3}{2}n$	$\tfrac{3}{2}(n+1)$
$\rho=5$	13	$\tfrac{8}{5}n$	$\tfrac{9}{5}n$ (see Ex. 13(e))	$\tfrac{3}{2}n$	$2n+2$
$\rho=6$	18	$\tfrac{5}{3}n$	$2n$ (see Ex. 13(e))	$2n$	$\tfrac{1}{2}(5n+3)$

Exercises.

13a For each of the critical graphs in Figure 14.4 (in the following
     chapter), verify

     (i)   the upper bound given in Theorem 13.1;

     (ii) the lower bounds given in the above table.

13b Use the result of Theorem 13.5 to prove that if G is a ρ-critical
     graph with n_j vertices of valency j, for $j = 2,3,\dots,\rho$, then for
     each k satisfying $2 \leq k \leq \rho-1$, we have

     $$\sum_{j=2}^{k} n_j \leq \tfrac{1}{2}(k - 1)n_\rho.$$

13c Without using the result of Theorem 13.6, prove that if G is a
     Hamiltonian ρ-critical graph of order n, then G has at least
     $\tfrac{1}{4}(n - 1)(\rho + 2)$ edges.

13d Prove that if Vizing's Conjecture is correct, then every planar graph
     with $\rho \geq 7$ is necessarily of class one.

13e Use the method of Theorem 13.4 to verify the entries $\tfrac{9}{5}n$ (for $\rho=5$) and
     $2n$ (for $\rho=6$) in the above table.   (Fiorini [68].)

14 Critical graphs of small order

In the last four chapters we have been concerned with the structural properties of critical graphs in general. We now turn our attention to a more de-
tailed investigation of critical graphs of small order - in particular, to
the study of critical graphs with less than ten vertices. Some of this discussion will spill over into the next chapter, where we shall provide some
further evidence for the truth of the Critical Graph Conjecture, previously
mentioned in Chapter 11.

Critical Graphs of Order   n ≤ 6

We start our discussion by looking at critical graphs with at most six
vertices. These are easy to determine since Figure 6.1 illustrates all of
the connected class two graphs with at most six vertices, and we need only
check these eight graphs to find which of them are critical. It turns out
that exactly three of them are critical (see Figure 14.1), corresponding

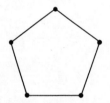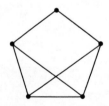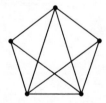

Fig. 14.1

respectively to ρ = 2, ρ = 3 and ρ = 4.

For later use we shall find it convenient to refer to critical graphs such
as these by means of a 'valency-list', in which we record the number of vertices of each valency. For example, the valency-lists of the three critical
graphs in Figure 14.1 are 2^5, 23^4 and 3^24^3, corresponding to the fact that
the first graph has five vertices of valency 2, the second has one vertex of

valency 2 and four of valency 3, and the third has two vertices of valency 3 and three of valency 4. More generally, if G is a graph with f_1 vertices of valency a_1, f_2 vertices of valency a_2, ... , and f_k vertices of valency a_k, where the a_i are arranged in increasing order, then the <u>valency-list</u> of G is the expression

$$a_1^{f_1} a_2^{f_2} \cdots a_k^{f_k} .$$

Critical Graphs of Order 7

Our next aim is to determine the valency-lists of all critical graphs of order 7. The argument we shall use is essentially a case-by-case analysis of the various possibilities for the minimum and maximum valencies of the graph. In each case, we start by using Corollary 11.3 and Theorems 13.1 and 13.5 to determine all of the possible valency-lists. In several cases we can then deduce that the corresponding graphs are ρ-critical because they are of class two (by Theorem 6.1), and contain no other ρ-critical subgraph; for simplicity, an argument of this kind will be called a <u>critical-list argument</u>. The following theorem is due to Beineke and Fiorini [14].

<u>Theorem 14.1</u>. A connected graph of order 7 is critical if and only if its valency-list is 2^7, 23^6, 24^6, $3^2 4^5$, 25^6, 345^5, $4^3 5^4$, $45^2 6^4$ or $5^4 6^3$.

<u>Proof</u>. Let G be a critical graph of order 7 with maximum valency ρ and minimum valency σ. In order to find the possible valency-lists for G, we shall consider all possible values of ρ and σ satisfying $2 \leq \sigma < \rho \leq 6$, together with the trivial regular case $\rho = \sigma = 2$. In the following, the number of vertices of valency k will be denoted by n_k, and the number of edges will be denoted by m.

$\underline{\rho = 2, \sigma = 2}$: The 7-circuit C_7 is the only connected graph of order 7 with $\rho = \sigma = 2$; it is clearly a 2-critical graph with valency-list 2^7.

$\underline{\rho = 3, \sigma = 2}$: By Theorem 13.5 we have $n_3 \geq 2n_2$, so that the only possible valency-list is 23^6. (Here, and in what follows, we frequently use the fact that there must be an even number of vertices of odd valency.) It then follows by a critical-list argument that all graphs with this valency-list must be critical.

$\underline{\rho = 4, \sigma = 2}$: By Theorems 13.1 and 13.5 we have $m \leq 13$ and $n_4 \geq n_3 + 2n_2$, so that the only possible valency-lists are $2^2 4^5$, $23^2 4^4$ and 24^6. Any

93

graph with valency-list $2^2 4^5$ can be obtained from the only regular 4-valent graph of order 6 by splitting one vertex of valency 4 into two of valency 2; since the original graph was 4-colourable, so is the resulting graph, so that $2^2 4^5$ cannot correspond to a critical graph. By using Vizing's Adjacency Lemma, one can easily prove that the only possibility for a critical graph with valency-list $23^2 4^4$ is the graph shown in Figure 14.2;

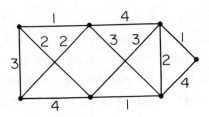

Fig. 14.2

but this graph is 4-colourable, so that $23^2 4^4$ cannot correspond to a critical graph. That every graph with valency-list 24^6 is critical follows by a critical-list argument.

$\rho = 4, \sigma = 3$: By Theorem 13.5 we have $n_4 \geq n_3$, so that the only possible valency-list is $3^2 4^5$. That every graph with this valency-list is critical follows by a critical-list argument.

$\rho = 5, \sigma = 2$: By Corollary 11.3 we have $n_5 \geq 5$, so that the only possible valency-lists are 235^5 and 25^6. In any graph with valency-list 235^5, the vertices of valency 2 and 3 cannot be adjacent, and hence, since K_6 is 5-colourable, so is this graph; so there are no critical graphs with valency-list 235^5. That every graph with valency-list 25^6 is critical follows by a critical-list argument.

$\rho = 5, \sigma = 3$: By Corollary 11.3, we have $n_5 \geq 4$, so that the only possible valency-lists are $3^2 45^4$ and 345^5. In any graph with valency-list $3^2 45^5$, the four vertices of valency 5 must all be adjacent to each other, and since the two vertices of valency 3 cannot be adjacent, the vertex of valency 4 must be adjacent to one vertex of valency 3; but there is only one graph satisfying these conditions (see Figure 14.3), and it is clearly 5-colourable, so there are no critical graphs with valency-list $3^2 45^5$. That every graph with valency-list 345^5 is critical follows by a critical-list

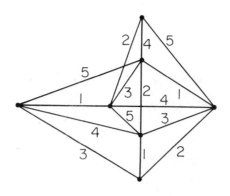

Fig. 14.3

argument.

$\rho = 5$, $\sigma = 4$: By Corollary 11.3 and Theorem 13.1 we have $n_5 \geq 3$ and $m \leq 16$, so that the only possible valency-list is $4^3 5^4$. That every graph with valency-list $4^3 5^4$ is critical follows by a critical-list argument.

$\rho = 6$, $\sigma = 2$: By Corollary 11.3 we have $n_6 \geq 6$, which is impossible-

$\rho = 6$, $\sigma = 3$: By Corollary 11.3 and Theorem 13.1 we have $n_6 \geq 5$ and $m \leq 19$, which are incompatible.

$\rho = 6$, $\sigma = 4$: By Corollary 11.3 and Theorem 13.1 we have $n_6 \geq 4$ and $m \leq 19$, so that the only possible valency-lists are $4^2 6^5$, $4^3 6^4$ and $45^2 6^4$. The first of these does not correspond to a graph, and the second corresponds to $K_7 - K_3$, which is easily seen to be 6-colourable. That every graph with valency-list $45^2 6^4$ is critical follows by a critical-list argument.

$\rho = 6$, $\sigma = 5$: By Corollary 11.3 and Theorem 13.1 we have $n_6 \geq 3$ and $m \leq 19$, so that the only possible valency-list is $5^4 6^3$. That every graph with valency-list $5^4 6^3$ is critical follows by a critical-list argument.

Since these are the only possibile cases, the proof is complete. □

By using the result of Theorem 14.1, Beineke and Fiorini [14] were able to determine all critical graphs of order 7 (see Figure 14.4); that they are all non-isomorphic may easily be verified by considering their complements.

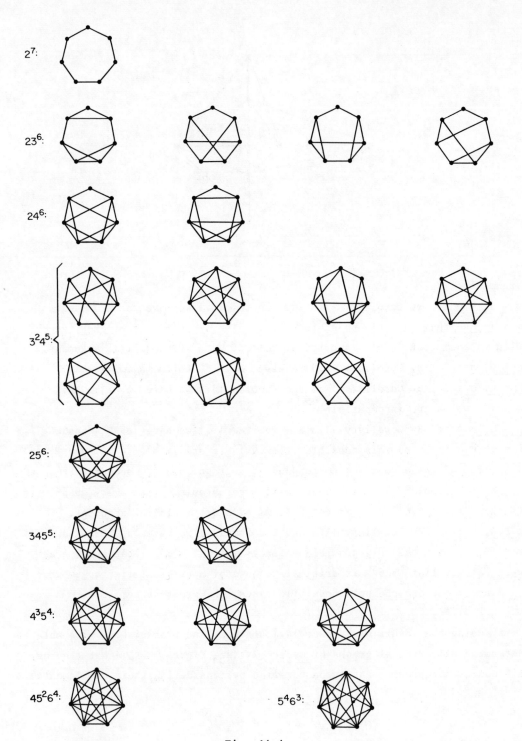

Fig. 14.4

Critical Multigraphs

It is natural to ask whether the case-by-case analysis used in Theorem 14.1 can be extended to multigraphs. In order to carry out such an analysis, one needs to have an analogue for multigraphs of Vizing's Adjacency Lemma. It seems likely that by combining the analogue given in Theorem 11.4 with a corresponding analogue of Theorem 13.5 one can use the method of Theorem 14.1. However, the technical details of such a procedure would probably be very cumbersome, and there has been little done in this direction.

If one restricts one's attention to 3-critical multigraphs, then the situation looks more promising. In [128], Jakobsen considered the problem of determining the 3-critical multigraphs of orders 4, 6 and 8, and was able to prove that no such multigraphs exist, as one might expect from the Critical Graph Conjecture. He also obtained all of the 3-critical multigraphs of orders 3, 5 and 7, and we conclude this chapter by drawing the reader's attention to Figure 14.5, where these multigraphs are presented.

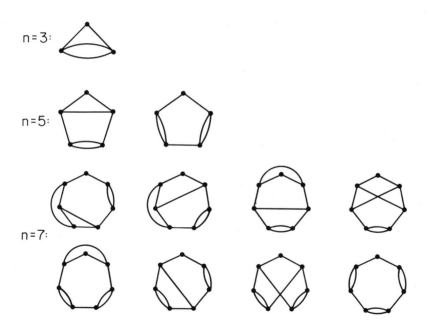

Fig. 14.5

Exercises.

14a (i)   Verify that if G is a ρ-critical graph of order 5, then G
           has exactly $2\rho + 1$ edges, and that if G is a ρ-critical graph
           of order 7, then G has exactly $3\rho + 1$ edges.

     (ii) By finding two 3-critical graphs of order 9 with different
           numbers of edges, show that there is no analogue of part (i) for
           ρ-critical graphs of order 9.

14b By using a case-by-case analysis, as in the proof of Theorem 14.1, give
     an alternative proof of the fact that there are no critical graphs of
     order 6.

14c Prove that no two of the graphs in Figure 14.4 are isomorphic.

14d Prove that if G is a connected regular graph of order 4, 6 or 8,
     then G is of class one.   (Fiorini [68].)

15 The critical graph conjecture

At the end of Chapter 11 we saw how Jakobsen's work on 3-critical graphs of small order led him to conjecture that critical graphs of even order do not exist. This conjecture is now known as the Critical Graph Conjecture, and remains unsettled. In support of this conjecture we have already shown that there are no critical graphs of order 4 or 6, and in Theorem 11.6 we saw that there are no 3-critical graphs of order 4, 6, 8 or 10. In this chapter we shall provide further evidence for the truth of the Critical Graph Conjecture by proving that there are no critical graphs of order 8 or 10, and no 3-critical graphs of order 12.

Critical Graphs of Order 8

We start this section by stating two results of Fiorini on the existence of 1-factors in critical graphs of small order. The proof of these results involves Tutte's theorem on the existence of a 1-factor in a graph (see [99, p.86]), and the reader is referred to [14,68] for further details.

Theorem 15.1. (i) If G is a critical graph of even order $n \le 10$, then G contains a 1-factor.

(ii) If G is a 3-critical graph of even order $n \le 26$, then G contains a 1-factor. $\square$

We can now prove the non-existence of critical graphs of order 8. As in the proof of Theorem 14.1, we shall make extensive use of Corollary 11.3 and of Theorems 13.1 and 13.5 in determining which valency-lists need to be considered in detail.

Theorem 15.2. There are no critical graphs of order 8.

Proof. Let G be a critical graph of order 8 with maximum valency ρ and minimum valency σ. Since G is not regular, by Theorem 10.5, and since the chromatic index of K_8 is 7, we may assume that $2 \le \sigma < \rho \le 6$. By Theorem 15.1(i), G contains a 1-factor F, and the graph $G-F$ contains

a $(\rho-1)$-critical subgraph H. We now consider all possible values of ρ
and σ satisfying $2 \le \sigma < \rho \le 6$. As before, the number of vertices of G
of valency k will be denoted by n_k, and the number of edges of G will
be denoted by m.

$\rho = 3, \sigma = 2$: By Theorems 13.1 and 13.5 we have $m \le 10$ and $n_3 \ge 2n_2$,
which are easily seen to be incompatible. So there is no critical graph in
this case.

$\rho = 4, \sigma = 2$: By Theorems 13.1 and 13.5 we have $m \le 13$ and $n_4 \ge n_3 + 2n_2$,
which are also incompatible.

$\rho = 4, \sigma = 3$: By Theorems 13.1 and 13.5 we have $m \le 14$ and $n_4 \ge n_3$, so
that the only possible valency-list is $3^4 4^4$. It follows that G-F must
have valency-list $2^4 3^4$, so that H cannot have exactly 7 vertices. So
H must have valency-list 23^4, and any extension to G results in three
vertices of valency 3 being mutually adjacent, which is impossible, by
Theorem 11.1.

$\rho = 5, \sigma = 2$: By Corollary 11.3 and Theorem 13.1 we have $n_5 \ge 5$ and
$m \le 16$, so that the only possible valency-list is $2^2 3 5^5$. By Vizing's
Adjacency Lemma (Theorem 11.2), such a valency-list cannot belong to a
critical graph, since otherwise four of the vertices of valency 5 would
have to be adjacent to a vertex of valency 2, and the others adjacent to
vertices of valency 5, which is impossible.

$\rho = 5, \sigma = 3$: By Corollary 11.3 and Theorem 13.1 we have $n_5 \ge 4$ and
$m \le 17$, so that the only possible valency-lists are $3^3 5^5$, $3^4 5^4$ and $3^2 4^2 5^4$.
It follows that G-F has valency-list $2^3 4^5$, $2^4 4^4$ or $2^2 3^2 4^4$, so that the
order of H cannot be 7, and must therefore be 5. But this means that H
must have valency-list $3^2 4^3$, and any extension to G requires two vertices
of valency 3 to be adjacent, which is impossible.

$\rho = 5, \sigma = 4$: By Theorems 13.1 and 13.5 we have $m \le 18$ and $3n_5 \ge 2n_4$, so
that the only possible valency-list is $4^4 5^4$. Moreover, H must have valency-list $3^2 4^3$, and cannot therefore be a subgraph of G-F which has valency-list $3^4 4^4$.

$\rho = 6, \sigma = 2$: By Corollary 11.3 and Theorem 13.1 we have $m \le 19$ and
$n_6 \ge 6$, which are incompatible.

$\rho = 6, \sigma = 3$: By Corollary 11.3 and Theorem 13.1 we have $m \le 20$ and
$n_6 \ge 5$, so that the only possible valency-list is $3^2 4 6^5$. It follows that
G-F has valency-list $2^2 3 5^5$ which cannot possibly admit a 5-critical sub-

100

graph of order 7.

$\rho = 6$, $\sigma = 4$: By Corollary 11.3 and Theorem 13.1 we have $m \leq 21$ and $n_6 \geq 4$, so that the only possible valency-lists are $4^4 6^4$, $4^3 6^5$ and $4^2 5^2 6^4$. The result now follows as in the previous case.

$\rho = 6$, $\sigma = 5$: By Corollary 11.3 and Theorem 13.1 we have $m \leq 22$ and $n_6 \geq 3$, so that the only possible valency-list is $5^4 6^4$. The result now follows as in the case $\rho = 6$, $\sigma = 3$.

Since these are the only possible cases, the proof is complete. ☐

Critical Graphs of Order 10

Using the result of Theorem 15.1(i), we can also prove the non-existence of critical graphs of order 10. Since some of the details are rather technical, we shall be content simply to outline the proof, referring the interested reader to Beineke and Fiorini [14] for a complete proof.

Theorem 15.3. There are no critical graphs of order 10.

Sketch of proof. Let G be a ρ-critical graph of order 10, with the smallest possible maximum valency ρ. By Theorem 15.1(i), G contains a 1-factor F, and the graph $G-F$ contains a $(\rho-1)$-critical subgraph H, which we assume to have maximum possible order. Using a valency-list argument it can be shown that the order of H must be exactly 9.

Let v be the vertex of G which is not in H, let k be its valency, and let σ be the minimum valency of G. Then the number of edges of $G-v$ satisfies

$$
\begin{aligned}
m(G - v) &= m(G) - k \\
&\leq (4\rho + \sigma - 1) - k \quad \text{(by Theorem 13.1)} \\
&\leq 4\rho - 1.
\end{aligned}
$$

It follows that $m(G - F - v) \leq 4\rho - 5$. Now there must be at least one vertex w of maximum valency adjacent to v in $G-F$, so that if t is the *second-largest* valency in H, then the valency of w lies between t and $\rho-2$. It follows that

$$
\begin{aligned}
m(H) &\leq m(G - F - v) - (\rho - 2 - t) \\
&\leq 3(\rho - 1) + t.
\end{aligned}
$$

But it can be proved by a rather technical valency-list argument that

$m(H) > 3(\rho - 1) + t$, thereby providing the required contradiction. □

3-Critical Graphs of Order 12

We conclude this chapter by proving that there are no 3-critical graphs of order 12. This extends the result of Theorem 11.6.

Theorem 15.4. There are no 3-critical graphs of order 12.
Proof. Let G be a 3-critical graph of order 12. By Theorem 13.5 G has at most four vertices of valency 2, and it follows from the result of Exercise 11d(ii) that G must have exactly four vertices of valency 2 and eight vertices of valency 3. Moreover, by Vizing's Adjacency Lemma, no vertex of valency 3 can be adjacent to two vertices of valency 2.

By Theorem 15.1(ii), G contains a 1-factor F, and the graph G-F is a graph of class two with valency list $1^4 2^8$. It follows that G-F must be the graph shown in Figure 15.1.

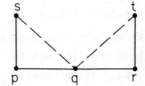

Fig. 15.1

We may assume that the pairs of vertices ad, bc, qs and qt are not adjacent in G, since otherwise G would be separable by two independent edges, and so, by Theorem 12.4, G would be a Hajós-union of two 3-critical graphs, one of which would have even order not exceeding 10, which is impossible.

Since every vertex of valency 3 must be adjacent in G to one of valency 2, it follows that a, b, p, q, r must generate a 5-circuit, and hence that one of the vertices v, w and z is adjacent to either s or t. If we assume, without loss of generality, that v is adjacent to s, then the edges vs, wz, ac, bd, rt and pq form a 1-factor F whose removal leaves a graph containing no odd circuits. But such a graph would be of class one, contradicting the fact that G-F must be of class two.

Exercises.

15a Prove Theorem 15.1.

15b Prove that all cubic bridgeless graphs of order 10 are of class one
     with the single exception of the Petersen graph.

15c In the proof of Theorem 15.3, prove that the subgraph H cannot have
     order 7.   (Beineke and Fiorini [14].)

15d Prove that there are no 3-critical graphs of order 14. (Andersen [2].)

Part IV—Further topics

In Part II of this book we described the Classification Problem arising from Vizing's Theorem, and in Part III we saw how the notion of critical graphs can be used to tackle this problem. We also took the opportunity to study some of the more important properties of critical graphs.

In this fourth and final part of our book we apply the methods and techniques developed earlier to some special classes of graphs. These include planar graphs (Chapter 16), and uniquely colourable graphs (Chapter 18). We also consider (in Chapter 17) some further properties of critical graphs relating to the length of circuits in such graphs. Finally, in Chapter 19 we define a parameter related to the chromatic index, called the 'multichromatic index'. It turns out that we can state analogues for this new parameter of the Four-Colour Theorem, the Critical Graph Conjecture, and several other results mentioned earlier in this book. We conclude by presenting proofs of these analogues, thereby providing extra supporting evidence for some of our earlier conjectures.

16 Planar graphs

At the end of Chapter 6, we stated a remarkable theorem asserting that every planar graph whose maximum valency is at least 8 is necessarily of class one. This result was first obtained by Vizing in 1965 [195], and one of the objects of this chapter is to present a proof of it. We shall also indicate various ways in which this theorem has been generalized, and we conclude with a result on the chromatic index of outerplanar graphs.

Vizing's Planar Graph Theorem

Before proving the above-mentioned theorem of Vizing, we shall present a weaker result which he obtained in an earlier paper [194]. Using the fact that every planar graph must contain a vertex whose valency is at most five, he proved the following result:

Theorem 16.1. If G is a planar graph whose maximum valency is at least 10, then G is of class one.

Proof. Suppose that G is a planar graph whose maximum valency ρ is at least 10, and suppose, without loss of generality, that G is ρ-critical. Since G is planar, there must be at least one vertex in G whose valency is at most five. If S is the set of all such vertices, then the subgraph of G induced by those vertices which are *not* in S is also planar, and must therefore contain a vertex (w, say) which is adjacent in G to at most five vertices which are not in S. But if v is a vertex in S to which w is adjacent in G, then the valency of v is at most five. It follows from Corollary 11.3 that w must be adjacent to at least $\rho - 4 > 5$ vertices of valency ρ, giving the required contradiction. □

We now prove the stronger form of Vizing's Planar Graph Theorem, which appeared in [195]; the proof we give is due to Mel'nikov [147].

Theorem 16.2. If G is a planar graph whose maximum valency is at least 8, then G is of class one.

Proof. Suppose that G is a planar graph whose maximum valency ρ is at least 8, and suppose, without loss of generality, that G is 8-critical.

Following the notation of Chapter 13, we let n_j be the number of vertices of valency j, for $2 \leq j \leq 8$, and let $n_8(r,s,t,u)$ be the number of vertices of valency 8 adjacent to exactly r vertices of valency 2, s vertices of valency 3, t vertices of valency 4, and u vertices of valency 5; we also let $n_7(s,t,u)$ be the number of vertices of valency 7 adjacent to s vertices of valency 3, t vertices of valency 4, and u vertices of valency 5.

By Theorem 3.3, we have

$$4n_2 + 3n_3 + 2n_4 + n_5 - n_7 - 2n_8 \geq 12. \qquad (\dagger)$$

We also have, by Vizing's Adjacency Lemma,

$$2n_2 = n_8(1,0,0,0)$$
$$3n_3 = n_8(0,1,0,0) + n_8(0,1,1,0) + n_8(0,1,0,1) + 2n_8(0,2,0,0) + n_7(1,0,0).$$

Also, since each vertex of valency 5 is adjacent to at least two vertices of valency 8, it follows from Theorem 11.2 that

$$2n_5 \leq n_8(0,1,0,1) + n_8(0,0,1,1) + n_8(0,0,0,1) + 2n_8(0,0,1,2) + n_8(0,1,0,0) + 2n_8(0,0,0,2) + 3n_8(0,0,0,3) + 4n_8(0,0,0,4).$$

But, by Theorem 10.1, each vertex of valency 4 can be adjacent only to vertices whose valency is at least 6, and by Theorem 11.2, each vertex of valency 4 adjacent to a vertex of valency 6 is also adjacent to three vertices of valency 8. It follows that if n_4' is the number of vertices of valency 4 adjacent to a vertex of valency 6, then we have

$$3n_4' + 4(n_4 - n_4') = n_8(0,1,1,0) + n_8(0,0,1,0) + n_8(0,0,1,1) + n_8(0,0,1,2) + 2n_8(0,0,2,0) + 2n_8(0,0,2,1) + 3n_8(0,0,3,0) + n_7(0,1,0) + n_7(0,1,1) + 2n_7(0,2,0).$$

Also, since each vertex of valency 4 is adjacent to at most two vertices

107

of valency 7, we have

$$n_4 - n_4' = n_7(0,2,0).$$

By combining all of these results, we get

$$4n_2 + 3n_3 + 2n_4 + n_5 \le n_7 + 2n_8,$$

contradicting the above result (†). This contradiction establishes the theorem. □

The Planar Graph Conjecture

One immediately asks what happens if the maximum valency of G is less than 8. It is not difficult to see that if the maximum valency of G is equal to 2, 3, 4 or 5, then G can lie either in class one or in class two. Examples of planar graphs of class two are the odd circuits ($\rho = 2$), and the graphs obtained by inserting a vertex into any edge of the graphs of the tetrahedron ($\rho = 3$), the octahedron ($\rho = 4$), and the icosahedron ($\rho = 5$).

However, the problem of determining what happens when the maximum valency is either 6 or 7 remains open. In this connection, the following conjecture was formulated by Vizing [195]:

Planar Graph Conjecture. If G is a planar graph whose maximum valency is at least 6, then G is of class one.

We saw in Exercise 13d that if Vizing's Conjecture on the number of edges of a critical graph is true, then every planar graph with maximum valency 7 is of class one. Further evidence in support of the Planar Graph Conjecture will be given in the final chapter of this book.

Results related to the Planar Graph Conjecture

Mel'nikov [147] was able to extend the results of Vizing to graphs embeddable in surfaces other than the sphere. In particular, he proved that if G is a graph which can be embedded in a surface with non-positive Euler characteristic η, and if its maximum valency ρ satisfies the inequality

$$\rho \geq \max \{[\tfrac{1}{2}(11 + \sqrt{(25 - 24\eta)})],[\tfrac{1}{3}(8 + 2\sqrt{(52 - 18\eta)})]\},$$

then G is of class one. The proof, which is similar to that of Theorem 16.2, will be omitted.

On the assumption that it is not easy to settle the Planar Graph Conjecture, we look for various restrictions on the graph which enable us to solve the problem at least partially. Using arguments similar to those in the proof of Theorem 16.2, it can be shown (see [68]) that if G is a planar graph of class two whose maximum valency is 7, then G has at least six vertices of valency 7. It follows that, if we impose restrictions on the number of vertices of maximum valency, then we can actually obtain positive results.

Similar restrictions can be made on the girth of planar graphs. In particular, Kronk, Radlowski and Franen [140] have stated the following results; their proofs may be found in [68]:

Theorem 16.3. Let G be a planar graph whose girth is g_0 and whose maximum valency is ρ_0; then G is of class one if any of the following conditions hold:

(i) $\rho_0 \geq 3$ and $g_0 \geq 8$;   (ii) $\rho_0 \geq 8$ and $g_0 \geq 3$;

(iii) $\rho_0 \geq 4$ and $g_0 \geq 5$;   (iv) $\rho_0 \geq 5$ and $g_0 \geq 4$. $\square$

Outerplanar Graphs

We conclude this chapter by showing that if we restrict ourselves to outerplanar graphs, then the Classification Problem is completely solved. We recall that a planar graph is outerplanar if it can be embedded in the plane in such a way that all its vertices lie on the boundary of the same face (see Figure 2.7). Although a proof of the following result can be found in [70], we present here a shorter version due to C. McDiarmid (private communication).

Theorem 16.4. An outerplanar graph G is of class one if and only if it is not a circuit of odd length.

Proof. If G is a circuit of odd length, then G is clearly an outerplanar graph of class two.

Conversely, let G be an outerplanar graph of class two. We may assume that its maximum valency ρ is at least 3 and that the order and maximum valency of G are chosen to be as small as possible, subject to these restrictions. Since the order of G is minimal, G must be 2-connected and hence Hamiltonian.

If G is of even order, then G has a 1-factor F consisting of alternate edges of the Hamiltonian circuit, so that G-F has maximum valency $\rho-1$. Thus, G-F is of class one, by our assumption on the minimality of ρ. It follows that G is of class one.

If, on the other hand, G is of odd order 2k+1, then G has a vertex v of valency 2, by outerplanarity. It follows that G has an independent set M of k alternate edges of the Hamiltonian circuit, covering all vertices except v. Clearly G-M is outerplanar and has maximum valency $\rho-1$, so that G-M is of class one, by our assumption on the minimality of ρ. It follows that G is also of class one, thereby completing the proof. □

Exercises.

16a If G is a given graph with maximum valency ρ, let $\lambda(G) = \max \sigma(G')$, where the maximum is taken over all induced subgraphs G' of G, and where $\sigma(G')$ is the minimum valency of G'. Prove that if $\lambda(G) \leq \frac{1}{2}\rho$, then G is of class one, and deduce the result of Theorem 16.1 as a special case.

16b Let G be a graph which can be embedded in the projective plane. If the maximum valency of G is at least 8, show that G is of class one.   (Mel'nikov [147].)

16c Show that if G is a planar graph of class two whose maximum valency is 6, then G has at least four vertices of maximum valency 6.

17 Circuit length in critical graphs

In this chapter we consider two problems related to the length of circuits in critical graphs. The first of these problems, discussed in the first section, deals with the girth (the length of any shortest circuit) of critical graphs. We showed in Chapter 16, how restrictions on the girth of graphs yield information on the Classification Problem for planar graphs. Here we show that ρ-critical graphs with arbitrary girth g exist for all values of ρ. We also discuss the related problem of determining $f(\rho,g)$, the minimum order of a ρ-critical graph of girth g.

In the second section, we consider problems relating to the circumference (the length of any longest circuit) of critical graphs. In particular, we prove a result of Vizing on the length of circuits in critical graphs, and conclude by obtaining a lower bound for the circumference of such graphs.

The Girth of Critical Graphs

The first natural question to ask about the girth of critical graphs is the analogue of that asked by Faber and Mycielski [66] and by Meredith [150] about class two graphs: do there exist ρ-critical graphs of arbitrary girth g for each ρ? If this is answered in the affirmative, one can then go on to ask: within what bounds can one expect to find ρ-critical graphs of given girth and of minimal order?

In order to answer the first question, we use various results on regular graphs. In particular, we shall need the following theorem of Sachs [162]:

<u>Theorem 17.1.</u> For each pair of integers ρ, g, satisfying $\rho \geq 3$ and $g \geq 2$, there exists a regular Hamiltonian graph G which is ρ-valent, has girth g, and in which all circuits of length g are mutually disjoint and constitute a 2-factor of G. $\square$

Once we have a ρ-valent graph G of the required girth g, it is easy to obtain from it a graph of class two with maximum valency ρ and girth g.

Now G is either of odd order or of even order. In the former case, G is itself of class two, by Corollary 6.3. In the latter case, the graph G' obtained from G by inserting a vertex into any edge of G is of class two, by Corollary 6.5. Having obtained a class two graph G' with maximum valency ρ and girth g, we can now consider a ρ-critical subgraph G" of G'; such a subgraph always exists, by Theorem 10.3, and its girth is not less than that of G'. It follows that for any integers ρ, g, satisfying $\rho \geq 3$ and $g \geq 3$, there exists a ρ-critical graph whose girth is at least g. (Recall that we asked you to prove part of this result in Exercise 6d(ii).) However, our aim is to sharpen this last statement to prove the existence of ρ-critical graphs whose girth is exactly equal to g. This is done as follows:

Theorem 17.2. For any integers ρ, g, satisfying $\rho \geq 3$ and $g \geq 3$, there exists a ρ-critical graph of girth g.
Proof. In order to prove this statement we use the regular graphs which Sachs obtained in order to establish Theorem 17.1. The ρ-valent graph $G(\rho,g)$ of girth g obtained by Sachs has the properties that (i) the Hamiltonian circuit H includes g-1 edges from each of the circuits of length g, (ii) $G(\rho,g)$ – H has $\rho-2$ 1-factors, and (iii) except for the case $\rho = 2$ and g odd, $G(\rho,g)$ has even order.

   Now consider the graph G' obtained from $G(\rho,g)$ by inserting a vertex into an edge of the Hamiltonian circuit which is incident to some vertex of a circuit Γ of length g, but does not itself belong to Γ. By Corollary 6.5, G' is of class two, and by Exercise 12b, G' is vertex-critical. Thus, if G" is any ρ-critical subgraph of G', then G" has the same order as G'. Moreover, the deletion of any edge of H reduces the chromatic index of G', since we can then colour the remaining edges of H with two colours and the edges of each of the $\rho-2$ 1-factors with a distinct colour. The same is also true for the edge of Γ which is not in H, since, by Vizing's Adjacency Lemma, a vertex of valency 2 can be adjacent only to vertices of maximum valency. It follows that no edge of Γ is deleted from G' to yield G", so that the girth of G" is exactly g, as required. □

   The second question proposed above seems difficult to answer in full generality. However, if we define $f(\rho,g)$ to be the minimal order of a ρ-critical graph of girth g, and if we restrict our attention to the cases g = 3

112

and g = 4, then the exact values of f(ρ,g) are as follows [72]:

Theorem 17.3. Let G be a ρ-critical graph of order n and girth g. Then

(i)   if g = 3, then n $\geq$ ρ+1, if ρ is even,

and n $\geq$ ρ+2, if ρ is odd;

(ii) if g = 4, then n $\geq$ 2ρ+1.

Moreover, there exist critical graphs which attain these bounds. □

For graphs of girth g $\geq$ 5, the exact value of f(ρ,g) is known only for some special cases. Thus, for example, it is known [72] that f(3,6) = 15. For other small values of ρ and g, only bounds on f(ρ,g) have been given. For example, it was shown in [68] that 21 $\leq$ f(3,7) $\leq$ 25, and that 15 $\leq$ f(4,5) $\leq$ 19. In each case, the method of proof consists in finding some critical graph with the required properties, thereby establishing the upper bound; the lower bound is then established by using properties of critical graphs to show that critical graphs of smaller order cannot exist.

In such arguments it would be very helpful if one could show that the function f(ρ,g) increases monotonically with respect to each variable; that is,

$$f(\rho,g) \leq f(\rho,g+1) \quad \text{and} \quad f(\rho,g) \leq f(\rho+1,g).$$

It is not difficult to show that the first inequality always implies the second, but unfortunately, the first inequality has not yet been established. We conjecture that it is valid.

The Circumference of Critical Graphs

In [195] Vizing used his Adjacency Lemma to obtain a bound on the circumference of a critical graph. In particular, he proved the following result:

Theorem 17.4. If G is a ρ-critical graph, then G contains a circuit whose length is at least ρ + 1.

Proof. Suppose on the contrary that every circuit of G has length at most ρ. Let C = a_1 $\rightarrow$ a_2 $\rightarrow$ ... $\rightarrow$ a_t $\rightarrow$ ... $\rightarrow$ a_s be a chain of maximum length in G such that a_1 is adjacent to a_t but to no vertex a_j with t < j $\leq$ s. Then the vertex a_1 can be adjacent only to other vertices in C, since otherwise we should get a longer chain. Also, a_1 cannot be of valency ρ, since otherwise $a_1 \to a_2 \to \ldots \to a_t \to a_1$ would be a circuit whose length is at least $\rho + 1$. It follows that the initial vertex in a chain of maximum length must have valency less than ρ and can be adjacent only to other vertices in the chain.

So let $\rho(a_1) = k$, where $k < \rho$. By Vizing's Adjacency Lemma, a_2 is adjacent to at least $\rho - k + 1$ vertices of valency ρ, and each such vertex must be in C, since otherwise, we should obtain a chain $a' \to a_2 \to \ldots \to a_s$ of maximum length in which the initial vertex a' had valency ρ.

Furthermore, if a_1 is adjacent to a_i, then $\rho(a_{i-1}) < \rho$, since otherwise the chain $a_{i-1} \to a_{i-2} \to \ldots \to a_1 \to a_i \to a_{i+1} \to \ldots \to a_s$ would be a chain of maximum length s in which the initial vertex had valency ρ.

It follows that among the vertices $\{a_1, \ldots, a_t\}$ there are $\rho - k + 1$ vertices of valency ρ, and k vertices whose valency is less than ρ. This implies that $a_1 \to a_2 \to \ldots \to a_t \to a_1$ is a circuit of length $(\rho - k + 1) + k = \rho + 1$, giving the required contradiction. $\square$

This result shows that $\rho + 1$ is a lower bound for the circumference of a ρ-critical graph. As with the case of Vizing's bound on the number of edges of a critical graph given in Theorem 13.2, this bound on the circumference does not involve the order of the graph, so that for graphs of large order it can be rather weak. To counteract this, we can make use of the fact that critical graphs are 2-connected (by Theorem 10.2), so that any critical graph with diameter d must contain a circuit whose length is at least $2d$. By obtaining bounds for the diameter of a critical graph of order n, we can then obtain the following bound for its circumference:

Theorem 17.5. If G is a ρ-critical graph of order n whose minimum valency is σ, then G contains a circuit whose length is at least

$$2 \frac{\log((n-1)(\rho-2)/\sigma)}{\log(\rho-1)}.$$

Proof. We shall make use of a standard argument in order to obtain a bound for the diameter of G. We split $V(G)$ into disjoint subsets $A_0, \ldots, A_t$ by letting A_0 consist of a single vertex v of valency σ, and letting A_j

(j = 1,2,...,t) consist of all those vertices whose distance from v is exactly j. It follows that

$$|A_0| = 1, \quad |A_j| \le \sigma(\rho - 1)^{j-1}, \quad \text{and} \quad n = \sum |A_i|,$$

giving

$$1 + \sigma + \sigma(\rho-1) + \ldots + \sigma(\rho-1)^{t-1} \ge n.$$

The result now follows by summing this geometric series, and using the fact that $t \le d$. □

Note that the bound given by this theorem is better than Vizing's bound if n is large. For example, if $\rho = 3$, then Theorem 17.4 gives 4 as a lower bound for the circumference, which is smaller than the bound of Theorem 17.5 if $n \ge 9$. More generally, it is easy to see that Theorem 17.5 gives a better estimate than Vizing's bound if n is greater than about ρ^ρ. However, it seems likely that the estimate of Theorem 17.5 can be improved further, and that a lower bound of order $\rho.2^t$, where $t = \dfrac{\log(2n)}{\log(2\rho)}$, can be obtained.

We conclude this chapter by stating an upper bound for the circumference of certain graphs. In [71], Fiorini described an explicit construction which yields an infinite family of critical graphs whose circumference can be estimated. In particular, he proved the following result:

Theorem 17.6. There exists an infinite family of ρ-critical graphs $\{G_k\}$ satisfying

$$c_k \le 4(\rho + 1)2^{q(k)},$$

where c_k is the circumference of G_k, n_k is the order of G_k, and

$$q(k) = \frac{\log n_k - \log(\rho-1)}{\log(2\rho - 2)} . \quad □$$

Exercises.

17a Show that there cannot exist

(i)   a 3-critical graph of girth seven whose order is at most 20;

(ii) a 4-critical graph of girth five whose order is at most 14.

17b (i)   Construct a 3-critical graph of girth six whose order is 15.

(ii) Construct a 3-critical graph of girth seven whose order is 25.

17c Prove that the inequality $f(\rho,g) \leq f(\rho,g+1)$ implies the inequality $f(\rho,g) \leq f(\rho+1,g)$.

18 Uniquely colourable graphs

In Exercise 3f we defined a graph G with chromatic number k to be uniquely colourable(v) if every k-colouring of the vertices of G induces the same
partition of the vertex-set into sets of vertices of the same colour. In
this chapter we shall look at the corresponding concept for edge-colourings,
and it will turn out that some rather unexpected results can be obtained.

Uniquely Colourable Graphs

If G is a graph with chromatic index k, then G is said to be
uniquely k-colourable (or, simply, uniquely colourable) if every k-colouring
of the edges of G induces the same partition of the edge-set into colour-
classes. For example, all circuits of even length and all open chains are
uniquely 2-colourable, the graphs K_3, K_4, K_4-e, and $K_4-\{e,e'\}$ (where e
and e' are edges) are all uniquely 3-colourable, and for each t, the star
graph $K_{1,t}$ is uniquely t-colourable.

In proving results on uniquely colourable graphs, it is frequently necessary to use the following elementary result:

Theorem 18.1. If G is a uniquely k-colourable graph, then each edge of G
is adjacent to edges of every other colour.

Proof. Let the colours used to colour G be denoted by $\alpha_1,\ldots,\alpha_k$, and
suppose that e is an edge with colour α_i which is not adjacent to any
edge coloured α_j, for some j. Then $H(\alpha_i,\alpha_j)$, the subgraph induced by all
those edges coloured α_i or α_j, is a disconnected graph. We can therefore
interchange the colours of the edges in any one of the components of
$H(\alpha_i,\alpha_j)$, to give a different partition into colour-classes. This contradicts the fact that G is uniquely colourable, thereby proving the result. ☐

It follows immediately from Theorem 18.1 that each subgraph of the form
$H(\alpha_i,\alpha_j)$ must be either a circuit of even length or an open chain. This in
turn shows that the colour-classes must differ in size by at most one, there-

by verifying the result of de Werra and McDiarmid (see Chapter 9) in the case of uniquely colourable graphs.

Uniquely k-Colourable Graphs, for k ≥ 4

It was conjectured in [69,201] that if k ≥ 4, then the only uniquely k-colourable graph is the star-graph $K_{1,k}$. This conjecture has recently been proved by A. Thomason [177], and we shall present a sketch of his proof. Note that it is sufficient to prove the result only in the case k = 4, since if G (≠ $K_{1,k}$) is any uniquely k-colourable graph with k ≥ 5, then the subgraph of G induced by those edges coloured with any four of the colours must be a uniquely 4-colourable graph.

We start by stating a result of Thomason [177] on the decomposition of a regular 4-valent graph or multigraph into two disjoint Hamiltonian circuits; such a pair of circuits is called a <u>Hamiltonian pair</u>.

<u>Theorem 18.2</u>. Let G be a regular 4-valent graph or multigraph with at least three vertices. If G has a Hamiltonian pair, then it has at least four such pairs.   ☐

Using the result of Theorem 18.1, Thomason showed that if there exists a uniquely 4-colourable graph G other than $K_{1,4}$, then there must exist a uniquely 4-colourable multigraph H which is either a 4-valent graph, or has the property that all but two of the vertices of H have valency 4. In this latter case, the two remaining vertices are of valency 2 and their incident edges must be coloured with the same pair of colours. Using these results, we can now present the main part of Thomason's proof:

<u>Theorem 18.3</u>. If G is a uniquely 4-colourable graph, then G = $K_{1,4}$.
<u>Proof</u>. Let G be a uniquely 4-colourable graph other than $K_{1,4}$. By virtue of the above remarks, we can assume that G is either a 4-valent graph, or that G has exactly two vertices of valency 2 and the rest of valency 4.

In the first case, the union of any two of the four colour-classes induces a Hamiltonian pair in G, so that G must have exactly three Hamiltonian pairs induced by these colour-classes. But by Theorem 18.2, G has at least four such pairs, and so G cannot be uniquely colourable.

In the second case, G has two vertices v and w of valency 2 with the same pair of colours appearing at each. By deleting v and w, and joining their neighbours appropriately, we can obtain from G another graph G' which also contradicts Theorem 18.2. Thus in either case the proof is complete. □

We conclude that uniquely k-colourable graphs are completely characterized when k ≥ 4. Since it is clear that the only uniquely 2-colourable graphs are the circuits of even length and the open chains, there remains only the case k = 3 to consider.

Uniquely 3-Colourable Graphs

We have already seen a few examples of uniquely 3-colourable graphs, such as K_3, K_4, and $K_{1,3}$. Further examples of such graphs can easily be obtained by taking any uniquely 3-colourable graph G (other than K_3) and replacing any vertex of valency 3 in G by a triangle. So, for example, if we start with K_4, we can take any vertex v and replace it by a triangle T; we can then choose any other vertex, and replace it by a triangle, and so on. This process is illustrated in Figure 18.1, and can clearly be used to gener-

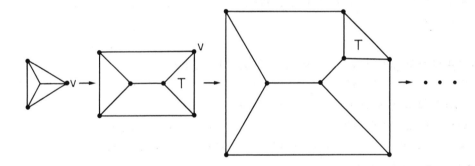

Fig. 18.1

ate an infinite family of uniquely 3-colourable cubic planar graphs. Consideration of these graphs leads one to make the following conjecture:

Conjecture 1. Every uniquely 3-colourable cubic planar graph contains a triangle.

Note that this conjecture becomes false if the word "planar" is omitted; for

119

example, the generalized Petersen graph $P(9,2)$ shown in Figure 7.2 is a non-planar uniquely 3-colourable cubic graph which contains no triangle. Note, however, that if this conjecture is true, then we can characterize uniquely 3-colourable cubic planar graphs as precisely those graphs obtained from K_4 by a sequence of transformations, each of which replaces a vertex by a triangle. By checking over the computer-generated list of cubic graphs of order at most 14, obtained by Bussemaker et $al.$ [36], it is easy to see that the order of any counterexample to Conjecture 1 must be at least 16.

Another conjecture on uniquely 3-colourable graphs is due to Greenwell and Kronk [88], and relates to the number of Hamiltonian circuits in such a graph. (Note that this echoes Thomason's approach to the problem.)

Conjecture 2. If G is a cubic graph with exactly three Hamiltonian circuits, then G is uniquely 3-colourable.

It is clear that every uniquely 3-colourable graph has exactly three Hamiltonian circuits. Since every bipartite cubic graph can be shown to have an even number of Hamiltonian circuits, it follows that every uniquely 3-colourable graph must contain an odd circuit – a result of some relevance to Conjecture 1.

The Chromatic Index of Uniquely Colourable Graphs

We conclude this chapter by proving that, with only one exception, uniquely k-colourable graphs are necessarily of class one. This was first proved by Greenwell and Kronk [88] and was later obtained independently by Fiorini [69]. In view of the above results of Thomason, we may restrict our attention to the case $k = 3$.

Theorem 18.4. If $G \neq K_3$ and G is uniquely 3-colourable, then G is of class one.

Proof. Let G be a uniquely 3-colourable graph of class two (other than K_3), and let the edges of G be coloured with the colours α, β, γ and δ. Let v_0 be a vertex of valency 3 incident to the edges v_0v_1, v_0v_2, v_0v_3 coloured α, β and γ respectively. Then the subgraphs induced by those edges coloured α and δ, β and δ, and γ and δ must all be open chains with initial vertex v_0. Since each of the vertices v_1, v_2 and v_3 is incident to some edge coloured δ, by Theorem 18.1, it follows that each of the two-coloured chains $H(\alpha,\delta)$, $H(\beta,\delta)$ and $H(\gamma,\delta)$ has at least four vertices.

If the chain $H(\alpha,\delta)$ has $\lambda + 1$ vertices, where $\lambda \geq 4$ (see Figure 18.2),

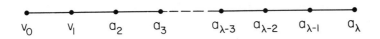

$$v_0 \quad v_1 \quad a_2 \quad a_3 \quad \cdots \quad a_{\lambda-3} \quad a_{\lambda-2} \quad a_{\lambda-1} \quad a_\lambda$$

Fig. 18.2

then some vertex x in the set $\{v_1, a_2, a_3, \ldots, a_{\lambda-1}\}$ is not an end-vertex in either $H(\beta,\delta)$ or $H(\gamma,\delta)$. So x must have valency 4, which is clearly a contradiction.

It follows that each of the subgraphs $H(\alpha,\delta)$, $H(\beta,\delta)$ and $H(\gamma,\delta)$ has exactly four vertices, namely v_0, v_1, v_2 and v_3. Hence G is of class one, giving the required contradiction. $\square$

Exercises.

18a Show that, up to isomorphism, the graph $P(9,2)$ of Figure 7.2 has exactly four 1-factors, and deduce that $P(9,2)$ is uniquely 3-colourable.

18b Let G be a uniquely 3-colourable graph with m edges and let $q = [\frac{1}{3}m]$ and $r = m - 3q$. Prove that G has $3 - r$ colour-classes of size q, and r colour-classes of size $q + 1$.

18c (i) Prove that if G is a uniquely 3-colourable graph of order n, then the number of edges of G is equal to $\frac{3}{2}n - r$, where $r = 0, 1, 2$ or 3.

(ii) Give examples to show that there exist uniquely 3-colourable graphs corresponding to each of these values of r.

18d Let G be a uniquely k-colourable graph, and let $e = vw$ be an edge of G. Suppose that we are given a k-colouring of the edges of G, and that Θ_v and Θ_w denote the sets of colours which appear at v and w respectively. Prove that

(i) $|\Theta_v \cup \Theta_w| = k;$

(ii) $|\Theta_v \cap \Theta_w| = \rho(v) + \rho(w) - k;$

(iii) $|\Theta_v \setminus \Theta_w| = k - \rho(w);$

(iv) $|\Theta_w \setminus \Theta_v| = k - \rho(v).$

(Compare these results with those of Exercise 10e)

18e Show that if G is a regular uniquely 3-colourable graph, other than K_3, then any graph H obtained from G by inserting a vertex into some edge of G is 3-critical.

19 The multichromatic index

In this final chapter, we shall consider a parameter which is related to the chromatic index. Although this parameter is more difficult to define than the chromatic index, it turns out that the analogues of some difficult or unsolved questions for the chromatic index are completely solved for this parameter. In particular, we shall prove the analogues of the following:

(1) The Four-Colour Theorem (Chapter 3);

(2) The Critical Graph Conjecture (Chapter 11);

(3) The Planar Graph Conjecture (Chapter 16).

These results strengthen our belief in the truth of the Critical Graph Conjecture and the Planar Graph Conjecture for the chromatic index.

The Multichromatic Index

If G is a graph, we denote by G^k the multigraph obtained from G by replacing each edge by a set of k multiple edges. The <u>multichromatic index</u> $\chi^*(G)$ of G is then defined by

$$\chi^*(G) = \inf \frac{1}{k} \chi'(G^k),$$

where the infimum is taken over all positive integers k. It can be shown that the infimum is actually attained - that is, that there is some integer k for which $\chi^*(G) = \frac{1}{k} \chi'(G^k)$.

The multichromatic index has been studied by several authors, notably Berge [18], Hilton [112], Seymour [165] and Stahl [167]. In order to discuss its properties, we shall need the following parameters. If H is a subgraph of G with $n(H)$ vertices and $m(H)$ edges, then we define

$$\tau(H) = \frac{2m(H)}{n(H)-1}, \quad \text{and} \quad \Phi(G) = \max \{\tau(H)\},$$

where the maximum is taken over all subgraphs H of G with odd order, and where the curly brackets have the meaning given to them at the end of Chapter 2. (It is not difficult to see that, for a graph G, $\Phi(G)$ is the same as the parameter defined in Exercise 5g.)

The first fundamental result on the multichromatic index is a consequence of a difficult theorem on matchings due to Edmonds [61]. Proofs of Theorem 19.1 may be found in [165,167].

Theorem 19.1. If G is a graph with maximum valency ρ, then
$$\chi^*(G) = \max\{\rho, \Phi(G)\}. \quad \square$$

The Analogue of the Planar Graph Conjecture

In this section, we shall show how the result of Theorem 19.1 can be used to give a simple proof of the multichromatic analogue of the Planar Graph Conjecture. The following result is due to Stahl [167]:

Theorem 19.2. If G is a planar graph whose maximum valency ρ is at least 6, then $\chi^*(G) = \rho$.

Proof. It follows from Euler's polyhedral formula that G has at most $3n-6$ edges, where n is the order of G. It follows that $\tau(G) \leq 6(n-2)/(n-1)$. But if H is any subgraph of G with n' vertices and m' edges, then H is a planar graph with $n' \leq n$. It follows that

$$\tau(H) = \frac{2m'}{n'-1} \leq \frac{6(n'-2)}{n'-1} \leq \frac{6(n-2)}{n-1},$$

and hence that

$$\Phi(G) \leq \frac{6(n-2)}{n-1} < 6.$$

The required result now follows from Theorem 19.1. $\quad \square$

The Analogue of the Critical Graph Conjecture

We next show how the result of Theorem 19.1 can be used to give a simple proof of the multichromatic analogue of the Critical Graph Conjecture. In this context, we define a connected graph to be critical* if, for each edge e of G, $\chi^*(G - e) < \chi^*(G)$. The results in this section are due to Stahl [167,168]:

Theorem 19.3. Every critical* graph other than K_2 has an odd number of vertices.

Proof. Let G ($\neq K_2$) be a critical* graph with maximum valency $\rho(G)$. If $\chi^*(G) = \rho(G)$, then, for each edge of G, we have

$$\rho(G) = \chi^*(G) > \chi^*(G - e) \geq \rho(G - e),$$

so that $\rho(G) > \rho(G - e)$ for each edge of G, which is impossible, since $G \neq K_2$.

It follows that $\chi^*(G) > \rho(G)$, and hence, by Theorem 19.1, that $\chi^*(G) = \Phi(G)$. It follows that there exists an induced subgraph H of odd order for which $\chi^*(G) = \tau(H)$. If $G \neq H$, then there exists an edge e, which lies in G but not in H, for which

$$\chi^*(G - e) \geq \tau(H) = \chi^*(G),$$

contradicting the fact that G is critical*. It follows that $G = H$, and hence that G has odd order. $\square$

This result can easily be strengthened to give a complete characterization of critical* graphs:

Theorem 19.4. Let G ($\neq K_2$) be a connected graph with maximum valency ρ. Then G is critical* if and only if G has odd order and, for each edge e of G, $\Phi(G) > \max \{\rho, \tau(G-e)\}$.

Proof. If G is critical*, then by the proof of Theorem 19.3, G has odd order and $\chi^*(G) > \rho$. Hence, by Theorem 19.1,

$$\Phi(G) = \chi^*(G) > \rho, \quad \text{and} \quad \Phi(G) = \chi^*(G) > \chi^*(G-e) \geq \tau(G-e),$$

for each edge e of G.

Conversely, if G has odd order and $\Phi(G) > \max \{\rho, \tau(G-e)\}$, then by Theorem 19.1,

$$\chi^*(G) = \Phi(G) > \max \{\rho, \tau(G-e)\} \geq \max \{\rho(G-e), \tau(G-e)\},$$

so that $\chi^*(G) > \chi^*(G-e)$, as required. $\square$

An analogue of the Four-Colour Theorem

We conclude by using Theorem 19.1 to prove a multichromatic extension of the four-colour theorem. By Tait's Theorem (Chapter 4) we may assume that the four-colour theorem is expressed in its edge-colouring form - namely, that every cubic map has chromatic index 3. Berge [18], Hilton [112] and Stahl [167] have proved the following related result:

Theorem 19.5. If G is a cubic bridgeless graph, then $\chi^*(G) = 3$.
Proof. We assume first that G is 3-connected. In this case, let H be any subgraph of G of odd order, and let K be the subgraph of H induced by those vertices which have valency 3 in H. If K is empty, then $\rho(H) \le 2$, and $\tau(H) \le 3$. If K is not empty, then $H-K$ separates K from $G-H$. Since G is 3-connected, it follows that H must have at least three vertices more than K, and hence that

$$2m(H) = \sum_{v \in V(H)} \rho(v) \le 3n(H) - 3.$$

So we have $\tau(H) \le 3$ in this case also. Since $\tau(H) \le 3$ for every subgraph H of odd order, we have $\Phi(G) \le 3$, and the result follows from Theorem 19.1.

So we may assume that G is not 3-connected. It follows that G has a separating set consisting of two vertices, and it follows (since G is cubic) that G has a cutset consisting of just two edges $v_0 v_1$ and $w_0 w_1$. We now prove the result by induction on the order of G. To start the proof, we observe that the smallest cubic graph is K_4, and $\chi^*(K_4) = 3$. So let G be a cubic 2-connected graph of smallest possible order for which $\chi^*(G) > 3$. Then, by Theorem 19.1, there must exist an induced subgraph H of smallest possible odd order for which $\tau(H) > 3$. We shall derive various properties of H, and use them to deduce that H cannot exist.

(i) We note first that H is connected. For if this were not the case, then there would exist graphs H_1 and H_2 such that $H_1 \cup H_2 = H$, $H_1 \cap H_2 = \emptyset$, and H_1 is of odd order. We should then have $2m(H_1) \le 3(n(H_1)-1)$, since H was of minimal order subject to $\tau(H) > 3$. We also have $2m(H_2) \le 3n(H_2)$, since $\rho(H_2) \le 3$. It follows that

$$2m(H) = 2m(H_1) + 2m(H_2) \le 3n(H_1) + 3n(H_2) - 3 = 3n(H) - 3,$$

contradicting the definition of H.

(ii) We next note that H is 2-connected. For if this were not the case, then there would exist graphs H_1 and H_2 such that $H_1 \cup H_2 = H$, and $H_1 \cap H_2 = \{v\}$, for some vertex v. We then have

$$2m(H) = 2m(H_1) + 2m(H_2) \le 3n(H_1) + 3n(H_2) = 3n(H) - 3,$$

which again contradicts the definition of H.

Without loss of generality, we can assume that v_1 and w_1 lie in the same component of $G-v_0-w_0$. Now, either $v_1 w_1$ is an edge of G, or it is not. If it is, we let v_2 be the vertex adjacent to v_1 (other than v_0) and w_2 be the vertex adjacent to w_1 (other than w_0). Again, either $v_2 w_2$ is an edge of G, or it is not. If it is, we let v_3 be the vertex adjacent to v_2 (other than v_1), and w_3 be the vertex adjacent to w_2 (other than w_1). Repeating this process, we eventually obtain an induced ladder-like subgraph L (see Figure 19.1) in which all of the vertices

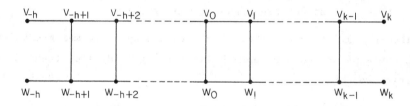

Fig. 19.1

except v_{-h}, w_{-h}, v_k and w_k have valency 3, and $v_{-h} w_{-h}$ and $v_k w_k$ are not edges of G.

Let G_1' and G_2' be the two components of $G - (L - \{v_{-h}, w_{-h}, v_k, w_k\})$, let G_1 be obtained from G_1' by including the edge $v_k w_k$, and let G_2 be obtained from G_2' by including the edge $v_{-h} w_{-h}$. Then G_1 and G_2 are both 2-connected cubic graphs.

By our inductive hypothesis, $V(H) \cap V(G_1) \ne \emptyset$, and $V(H) \cap V(G_2) \ne \emptyset$, so that $V(H) \supset V(L)$, since H is 2-connected. Now let H_1 be the subgraph of G_1 induced by $V(H) \cap V(G_1)$, and let H_2 be the subgraph of G_2 induced by $V(H) \cap V(G_2)$. Without loss of generality, we may assume that

H_1 is of odd order, and that H_2 is of even order. It follows that

$$2m(H_1) \leq 3n(H_1) - 3 \qquad \text{and} \qquad 2m(H_2) \leq 3n(H_2).$$

Also,

$$n(H) = n(H_1) + n(H_2) + 2(h + k - 1).$$

But

$$E(H) = (E(H_1) \setminus v_{-h}w_{-h}) \cup (E(H_2) \setminus v_k w_k) \cup E(L).$$

It follows that

$$
\begin{aligned}
2m(H) &= 2(m(H_1) - 1) + 2(m(H_2) - 1) + 2(3h + 3k - 1) \\
&\leq 3(n(H_1) - 1) + 3n(H_2) + 6(h + k - 1) \\
&= 3n(H) - 3,
\end{aligned}
$$

again contradicting the definition of H. This contradiction establishes the result. $\square$

Exercises.

19a Prove that

(i) $\chi^*(C_{2k+1}) = 2 + \dfrac{1}{k}$,

(ii) If P is the Petersen graph, then $\chi^*(P) = 3$.

19b Use Theorem 19.1 to prove that if G is a graph with maximum valency ρ which contains an odd subgraph H satisfying $\tau(H) > \rho$, then G is of class two. (Stahl [167].)

19c Let G be a graph with maximum valency ρ. Prove that

(i) $\chi^*(G) \leq \rho + 1$;

(ii) $\chi^*(G) = \rho + 1$ if and only if G is a complete graph of odd order.

(iii) If G is not a complete graph of odd order, then

$\chi^*(G) \leq \rho + \dfrac{\rho-1}{\rho+1}$, if ρ is odd,

and

$\chi^*(G) \leq \rho + \dfrac{\rho}{\rho+2}$, if ρ is even.

(iv) Give examples to show that the upper bounds in part (iii) can be attained for all values of ρ. (Stahl [168].)

19d Let G be a graph with m edges, and maximum valency ρ. Prove that if G is not a complete graph of odd order, then $\chi^*(G) \leq \dfrac{m}{m+1}(\rho + 1)$. (Hilton [107].)

128

19e Let G be a graph of order n with maximum valency ρ, and let G
have Euler characteristic η. Use the result of Theorem 19.1 to prove
that if $\rho \geq \dfrac{6(n - \eta)}{n - 1}$, then $\chi^*(G) = \rho$.   (Stahl [167].)

Guide to the bibliography

Our aim here is to survey the literature on edge-colourings of graphs by summarizing the bibliography which follows. The terminology used here is explained earlier in the book.

As we have seen in Chapter 5, the most important result in the theory of edge-colourings is Vizing's Theorem (Theorem 5.1) first published in 1964. Prior to this date, most of the early papers were concerned with the relationship between edge-colourings and vertex-colourings. These include [26,65,100,124,125,143,146,160,164,178,185-188]. Papers of a similar nature published subsequently include [7,169,183]. Proofs of Vizing's Theorem may be found in [12,17,27, 62,80,94,157,191,192,206].

As we noted in Chapter 6, Vizing's Theorem gives us a way of classifying graphs into two classes. The two principal lines of attack on this problem may be categorized as follows:

(i)   the classification of particular types of graph;
(ii) the use of critical graphs.

We consider each of these in turn.

(i) Regular graphs have received much attention in the literature. In particular, cubic graphs are studied in [36,120,138,139,153,161,163,165, 170-173,186-189]. Various generalizations of the Petersen graph (see Chapter 7) may be found in [23,25,42,44,49,82,119,131,149,150,197,198,205]. Other similar constructions not related to the Petersen graph are to be found in [15,66,71,76,127,173]. Most works of a general nature, such as [9,10,12,16,99,200,203,206] include a discussion of bipartite graphs and complete graphs. Further references on bipartite graphs include [6,32,33, 53,57, 79,93,95,135,136], and on complete graphs include [13,37-41,194].

The class of O_k graphs (see Chapter 7) is studied in [19,144,151,152]. [115] considers the Cartesian product of graphs, and [141] and [158] consider respectively complete r-partite graphs and generalized circuits. Pseudo-regular graphs (graphs whose valencies are either ρ or 1) are considered in [121-123], and [126,137] deal with line-graphs. Certain special graphs with remarkable properties are considered in [21,181]. The chromatic index of complementary graphs is discussed in [1,194].

(ii) Works of a general nature on critical graphs include [2,15,69,71, 72,76,77,110,194,195,202,203]. These include both the investigation of properties of critical graphs, and applications of critical graphs to the Classification Problem.

Bounds on the number of edges of critical graphs (Chapter 13) may be found in [15,68,69,71,76,77,195,196], whereas circuit length properties relating to the girth and the circumference of critical graphs (see Chapter 17) are discussed in [66,72,194] and [71,195] respectively. Hamiltonian graphs are considered in [69,88].

Using properties of critical graphs, it is shown in [64] that almost all graphs are of class one. Further applications of critical graphs include the investigation of graphs with particular topological properties [90,147,148, 170-173]. In particular, planar graphs (see Chapter 16) are discussed in [63,70,77,139,147,163,189,190].

Graphs of small order (Chapters 14,15) are considered in [14,15,128], whereas uniquely colourable graphs (Chapter 18) are investigated in [37,69, 73,88,177,201]. What evidence there is for the Critical Graph Conjecture (see Chapter 11) may be found in [14,15,69,167,201,204].

In addition to these two main lines of investigation, some attention has been given to the study of the distribution of colours in edge-coloured graphs. Works of this nature include [2-4,53,59,79,83,86,97,199]. Various relationships between edge-colourings and groups are considered in [20,37-39,116], whereas edge-colourings and Ramsey numbers are treated in [142, 184]. [11,132,159,163] discuss the number of ways of colouring a graph with a given number of colours.

Although simple graphs have received most attention in the literature, multigraphs have not been entirely ignored. Works of a general nature on multigraphs include [2,17,29,30,157,194]. Bounds on the chromatic index are discussed in [2,3,86,87,96,106,111,114,191,194,206], and critical multi-graphs are studied in [2,85,127-129]. Bipartite multigraphs are considered in [54,57,83,95,108], and other types of multigraph may be found in [3,4,

29,30,67,84].

Extensions of the theory of edge-colourings of graphs have been made to hypergraphs [28] and infinite graphs [28,29,31,48,92,155]. Equitable colorations of graphs (colourings in which the sizes of the colour-classes differ by at most one) are considered in [50,52,54-58], whereas good k-colourings (colourings in which the number of distinct colours at each vertex v is min $\{k,\rho(v)\}$) are studied in [56-58,80]. The achromatic index is discussed in [31], and total colourings are treated in [9,10,13].

Other related parameters include the cover index [80,91,96,108,109] and the multichromatic index [109,112,167].

Finally, applications of edge-colourings have been made to time-tabling problems [19,27,50,51,78,145,199], latin squares [38,39,78,96,113], networks [132], and matrix algebra [145]. Various relationships between edge-colourings and chemistry, quantum field theory, tensor algebra, and matroids are considered in [7,8], [154], [78,159] and [79] respectively.

(Note: In the following Bibliography, a reference of the form MR 50-9643 indicates review No. 9643 in Volume 50 of Mathematical Reviews.)

Bibliography

1   Y. Alavi and M. Behzad, Complementary graphs and edge chromatic numbers,
        Siam J. Appl. Math. $\underline{20}$ (1971) 161-163. MR 44-3923.

2   L.D. Andersen, Edge-colourings of simple and non-simple graphs, Aarhus
        University (1975).

3   L.D. Andersen, On edge-colourings of non-simple graphs, Pre-print Series
        No 21, Aarhus University (1976).

4   L.D. Andersen, On edge-colourings of graphs, Proc. Paris Conference
        (1976) (to appear).

5   K. Appel and W. Haken, Every planar map is four-colorable, Bull. Amer.
        Math. Soc. $\underline{82}$ (1976) 711- 712.

6   J.W. Archbold and C.A.B. Smith, An extension of a theorem of König on
        graphs, Mathematika $\underline{9}$ (1962) 9-10. MR 26-754.

7   A.T. Balaban, D. Francasiu and R. Banica, Graphs of multiple 1,2-shifts
        in carbonium ions and related systems, Rev. Roumaine
        Chim. $\underline{11}$ (1966) 1205-1227.

8   A.T. Balaban, Chemical graphs, Part XIII: Combinatorial Patterns, Rev.
        Roumaine Math. Pures et Appl. $\underline{17}$ (1972) 3-16.

9   M. Behzad, Graphs and their Chromatic Numbers, Doctoral Thesis, Michigan
        State University (1965).

10 M. Behzad and G. Chartrand, An introduction to total graphs, Theory of
        Graphs (Ed., P. Rosenstiehl) Dunod, Paris (1967) 31-33.

11 M. Behzad and G. Chartrand, The line-chromatic polynomial of a graph,
        Portugal Math. $\underline{27}$ (1968) 31-41. MR 41-5242.

12 M. Behzad and G. Chartrand, Introduction to the Theory of Graphs, Allyn
        and Bacon, Boston (1971).

13 M. Behzad, G. Chartrand and J.K. Cooper, The colour numbers of complete

graphs, J. London Math. Soc. <u>42</u> (1967) 226-228. MR 34-7396.

14 L.W. Beineke and S. Fiorini, On small graphs critical with respect to edge-colourings, Discrete Math. <u>16</u> (1976) 109-121.

15 L.W. Beineke and R.J. Wilson, On the edge-chromatic number of a graph, Discrete Math. <u>5</u> (1973) 15-20. MR 47-4836.

16 C. Berge, Graph theory, Amer. Math. Monthly <u>71</u> (1964) 471-481. MR 30-655.

17 C. Berge, Graphs and Hypergraphs, North-Holland, Amsterdam (1973). MR 52-5453.

18 C. Berge, The multicolorings of graphs and hypergraphs, Theory and Applications of Graphs, Springer Lecture Notes (to appear).

19 N.L. Biggs, An edge-colouring problem, Amer. Math. Monthly <u>79</u> (1972) 1018-1020.

20 N.L. Biggs, Pictures, Combinatorics (Eds., D.J.A. Welsh and D.R. Woodall) Inst. Math. Appl., Southend-on-Sea (1972) 1-17. MR 49-7168.

21 N.L. Biggs, Three remarkable graphs, Canad. J. Math. <u>25</u> (1973) 397-411. MR 48-156.

22 N.L. Biggs, E.K. Lloyd and R.J. Wilson, Graph Theory 1736-1936, Oxford Univ. Press (1976).

23 S. Bilinski and D. Blanuša, Proof of the indecomposability of a certain graph, Hrvatsko Prirodoslovno Društvo Glasnik Mat.-Fiz. Astr. Ser. II <u>4</u> (1949) 78-80. MR 11-377.

24 G.D. Birkhoff, The reducibility of maps, Amer. J. Math. <u>35</u> (1913) 115-128.

25 D. Blanuša, Problem ceteriju boja (The problem of four colours), Hrvatsko Prirodoslovno Društvo Glasnik Mat.-Fiz. Astr. Ser. II <u>1</u> (1946) 31-42. MR 10-136.

26 C. Blatter, On the algebra of the four-colour problem, Enseignment Math. (2) <u>11</u> (1965) 175-193. MR 32-516.

27 J.A. Bondy and U.S.R. Murty, Graph Theory with Applications, Elsevier,
        New York (1976).

28 R. Bonnet and P. Erdős, The chromatic index of an infinite complete
        hypergraph: A partition theorem, Hypergraph Seminar,
        Ohio State University, Columbus, Ohio, 1972. Springer
        Lecture Notes No. 411 (1974) 54-60. MR 51-10145.

29 J. Bosák, Chromatic index of finite and infinite graphs, Czech. Math. J.
        22 (1972) 272-290. MR 46-8862.

30 J. Bosák and J. Fiamčik, Chromatic index of Hamiltonian graphs, Mat.
        Cas. 23 (1973) 88-94. MR 48-1959.

31 J. Bosák and J. Nešetřil, Complete and pseudo-complete colourings of a
        graph (to appear).

32 A. Brace and D.E. Daykin, Pseudo-matchings of a bipartite graph, Proc.
        Amer. Math. Soc. 42 (1974) 28- 32. MR 48-8299.

33 A. Brace and D.E. Daykin, Pseudo-matchings of a bipartite graph,
        Corrigendum and addendum (to appear).

34 R.L. Brooks, On colouring the nodes of a network, Proc. Cambridge Phil.
        Soc. 37 (1941) 194-197. MR 6-281.

35 R.A. Brualdi, The chromatic index of the graph of the assignment
        polytope (to appear).

36 F.C. Bussemaker, S. Čobeljić, D.M. Cvetković, J.J. Seidel, Computer
        investigation of cubic graphs, J. Comb. Theory (B) (to
        appear).

37 P.J. Cameron, On groups of degree n and n-1, and highly symmetric
        edge-colourings, J. London Math. Soc. (2) 9 (1975) 385-
        391. MR 50-13217.

38 P.J. Cameron, Minimal edge-colourings of complete graphs, J. London
        Math. Soc. (2) 11 (1975) 337-346.

39 P.J. Cameron, Parallelisms of Complete Designs, L.M.S. Lecture Notes
        No. 23, Cambridge (1976).

40 P.J. Cameron, Embedding edge-colored complete graphs in binary affine
        spaces, J. Comb. Theory (A) <u>21</u> (1976) 203-215.

41 P.J. Cameron and J.H. van Lint, Graph Theory, Coding Theory and Block
        Designs, L.M.S. Lecture Notes No. 19, Cambridge (1975).

42 F. Castagna and G. Prins, Every generalized Petersen graph has a Tait
        colouring, Pacific J. Math. <u>40</u> (1972) 53-58. MR 46-3358.

43 A. Cayley, On the colouring of maps, Proc. Roy. Geog. Soc. (New Ser.) <u>1</u>
        (1879) 259-261.

44 G. Chartrand and J.B. Frechen, On the chromatic number of permutation
        graphs, Proof Techniques in Graph Theory (Ed., F. Harary)
        Academic Press, New York (1969) 21-24. MR 40-4165.

45 G. Chartrand and F. Harary. Planar permutation graphs, Ann. Inst. H.
        Poincaré (B) <u>3</u> (1967) 433-438. MR 37-2626.

46 H.S.M. Coxeter, Map-coloring problems, Scripta Math. <u>23</u> (1957) 11-25.
        MR 20-7277.

47 A. Cruse, On embedding incomplete symmetric latin squares, J. Comb.
        Theory (A) (to appear).

48 N. de Bruijn and P. Erdős, A colour problem for infinite graphs and a
        problem in the theory of relations, Kon. Ned. Akad. Wet.
        Proc. A, <u>54</u> (1951) 371-373. MR 13-763.

49 B. Descartes, Network colourings, Math. Gazette <u>32</u> (1948) 67-69. MR 10-
        136.

50 D. de Werra, On some combinatorial problems arising in scheduling, Canad.
        Oper. Res. Soc. J. <u>8</u> (1970) 165-175. MR 43-1878.

51 D. de Werra, Balanced schedules, INFOR – Canad. J. Oper. Res. and Inf.
        Proc. <u>9</u> (1971) 230-237. MR 45-1685.

52 D. de Werra, Equitable colorations of graphs, Rev. Fran. Inf. Rech. Opér.
        <u>5</u> (1971) 3-8.

53 D. de Werra, Investigations on an edge-coloring problem, Discrete Math.
        <u>1</u> (1972) 167-179. MR 45-3248.

54 D. de Werra, Decompositions of bipartite multigraphs into matchings,

Zeitschr. Oper. Res. 16 (1972) 85-90. MR 46-5182.

55 D. de Werra, A note on graph coloring, Rev. Fran. Aut. Inf. Rech. Opér.
Sér. Rouge 8 (1974) 49-53. MR 49-4841.

56 D. de Werra, On good and equitable edge-colorings, Cahiers Centre Études
Rech. Opér. 17 (1975) 417-426.

57 D. de Werra, An extension of bipartite multigraphs, Discrete Math. 14
(1976) 133-138.

58 D. de Werra, Some remarks on good colorations, J. Comb. Theory (B) 21
(1976) 57-64.

59 D. de Werra, On colour-feasible sequences of almost bipartite multigraphs, Proc. Paris Conf., 1976 (to appear).

60 G.A. Dirac, A property of 4-chromatic graphs and some remarks on
critical graphs, J. London Math. Soc. 27 (1952) 85-92.
MR 13-572.

61 J. Edmonds, Maximum matching and a polyhedron with (0,1) vertices, J.
Res. Nat. Bur. Standards 69B (1965) 125-130. MR 32-1012.

62 A. Ehrenfeucht and V. Faber, [A new proof of Vizing's theorem] (to
appear).

63 A. Ehrenfeucht, J.L. Hursch Jr., C. Morgenstern, Growth number and
colorability of graphs II, Notices Amer. Math. Soc. 23
(1976) A-512.

64 P. Erdős and R.J. Wilson, On the chromatic index of almost all graphs,
J. Comb. Theory (B) (to appear).

65 A. Errera, Sur un théorème de M. Whitney, un problème de Lebesgue et les
réseaux de Tait, III Congr. Nat. Sci. Bruxelles 2 (1950)
51-55. MR 17-69.

66 V. Faber and J. Mycielski, Graphs with valency k, edge-connectivity k,
chromatic index k+1 and arbitrary girth, Discrete Math.
4 (1973) 339-345. MR 47-4839.

67 J. Fiamčik and E. Jucovič, Colouring the edges of a multigraph, Arch.
Math. (Basel) 21 (1970) 446-448. MR 44-6546.

68 S. Fiorini, The Chromatic Index of Simple Graphs, Doctoral Thesis, The
        Open University, England (1974).

69 S. Fiorini, On the chromatic index of a graph, III: Uniquely edge-
        colourable graphs, Quart. J. Math.(Oxford) (3) $\underline{26}$ (1975)
        129-140. MR 51-7925.

70 S. Fiorini, On the chromatic index of outerplanar graphs, J. Comb.
        Theory (B) $\underline{18}$ (1975) 35-38. MR 71-2971.

71 S. Fiorini, Some remarks on a paper by Vizing on critical graphs, Math.
        Proc. Camb. Phil. Soc. $\underline{77}$ (1975) 475-483. MR 51-10146.

72 S. Fiorini, On the girth of graphs critical with respect to edge-
        colourings, Bull. London Math. Soc. $\underline{8}$ (1976) 81-86.

73 S. Fiorini, Un grafo cubico, non-planare, unicamente tricolorabile, di
        vita 5, Calcolo $\underline{13}$ (1976) 105-108.

74 S. Fiorini, A bibliographic survey of edge-colourings, J. Graph Theory
        (to appear).

75 S. Fiorini, Counterexamples to two conjectures of Hilton (to appear).

76 S. Fiorini and R.J. Wilson, On the chromatic index of a graph, I, Cahiers
        Centre Études Rech. Opér. $\underline{15}$ (1973) 253-262. MR 50-6864a.

77 S. Fiorini and R.J. Wilson, On the chromatic index of a graph, II,
        Combinatorics (Eds., T.P. McDonough and V.C. Mavron)
        L.M.S. Lecture Notes No. 13, Cambridge (1974) 37-51.
        MR 50-6864b.

78 S. Fiorini and R.J. Wilson, Edge-colourings of graphs - some applications,
        Proc. Fifth British Comb. Conf., 1975, Utilitas Math.,
        Winnipeg (1976) 193-202. MR 52-13461.

79 J. Folkman and D.R. Fulkerson, Edge-colorings in bipartite graphs,
        Combinatorial Math. and its Appl. (Eds., R.C. Bose and
        T.A. Dowling) Univ. N. Carolina Press, Chapel Hill (1969)
        561-577. MR 41-6722.

80 J.C. Fournier, Colorations des arêtes d'un graphe, Cahiers Centre Études
        Rech. Opér. $\underline{15}$ (1973) 311-314. MR 50-1952.

81 J.C. Fournier, Methode et théorème général de coloration des arêtes d'un

multigraphe, Proc. Paris Conf., 1976 (to appear).

82 M. Gardner, Mathematical games, Scientific American (1976) <u>234</u> No. 4,
126-130, and No. 9, 210-211.

83 D.P. Geller and A.J.W. Hilton, How to color the lines of a bigraph,
Networks <u>4</u> (1974) 281-282. MR 49-4834.

84 M.K. Gol'dberg, Нечетные циклы мультиграфа с большим хроматическим
классом (Odd cycles of a multigraph of large chromatic
class), Uprav. Syst. <u>10</u> (1972) 45-47.

85 M.K. Gol'dberg, О мультиграфах с хроматическим классом, близким к
максимальному (On multigraphs of almost maximal chromatic
class), Diskret. Analiz <u>23</u> (1973) 3-7. MR 50-6907.

86 M.K. Gol'dberg, Замечание о хроматическом классе мультиграфа (Remark on
the chromatic class of a multigraph), Vycislitel'naya
Matematika i Vycislitel'naya Tekhnika <u>5</u> (1974) 128-130.

87 M.K. Gol'dberg, Строение мультиграфов с ограничением на хроматический
класс (Structure of multigraphs with restrictions on the
chromatic class), Pre-print, Academy of Sciences of
Ukrainian SSR, Khar'kov (1976).

88 D. Greenwell and H. Kronk, Uniquely line-colorable graphs, Canad. Math.
Bull. <u>16</u> (1973) 525-529. MR 50-158.

89 H. Grötzsch, Zur Theorie der diskreten Gebilde. 14. Mitteilung: Ein
Kantensignierungssatz für Vierkantnetze auf der Kugel,
Wiss. Z. Martin-Luther Univ. Halle-Wittenberg Math. Nat.
Reihe <u>10</u> (1961/2) 843-850. MR 25-2595.

90 B. Grünbaum, Conjecture 6, Recent Progress in Combinatorics (Ed., W.T.
Tutte) Academic Press, New York (1969) 343.

91 R.P. Gupta, A theorem on the cover index of an s-graph, Notices Amer.
Math. Soc. <u>13</u> (1966) 714.

92 R.P. Gupta, The chromatic index and the degree of a graph, Notices Amer.
Math. Soc. <u>13</u> (1966) 719.

93 R.P. Gupta, A decomposition theorem for bipartite graphs, Théorie des
Graphes (Ed., P. Rosenstiehl) Dunod, Paris (1967) 135-136.

94 R.P. Gupta, Studies in the Theory of Graphs, Thesis, Tata Inst. Fund. Res., Bombay (1967).

95 R.P. Gupta, On decompositions of a multigraph into spanning subgraphs, Bull. Amer. Math. Soc. 80 (1974) 500-502. MR 49-149.

96 R.P. Gupta, On the chromatic index and the cover index of a multigraph (to appear).

97 R. Häggkvist, A partial solution of the Evans Conjecture for Latin squares, Umeå Univ. Pre-prints, No. 6 (1976).

98 G. Hajós, Über eine Konstruktion nicht n-färbarer Graphen, Wiss. Z. Martin-Luther Univ. Halle-Wittenberg Math. Nat. Reihe 10 (1961) 116-117.

99 F. Harary, Graph Theory, Addison-Wesley, Reading, Mass. (1969). MR 41-1566.

100 P.J. Heawood, Map-colour theorem, Quart. J. Pure Appl. Math. (Oxford) 24 (1890) 332-338.

101 P.J. Heawood, On the four-colour map theorem, Quart. J. Pure Appl. Math. (Oxford) 29 (1898) 270-285.

102 P.J. Heawood, On extended congruences connected with the four-colour map theorem, Proc. London Math. Soc. (2) 33 (1932) 253-286.

103 P.J. Heawood, Note on a correction in a paper on map-congruences, J. London Math. Soc. 19 (1944) 18-22. MR 6-165.

104 P.J. Heawood, Map-colour theorem, Proc. London Math. Soc. (2) 51 (1949) 161-175. MR 11-43.

105 H. Heesch, Untersuchungen zum Vierfarbenproblem, B. I. Hochschulskripten 810/810a/810b, Bibliographisches Institut, Mannheim/ Vienna/Zürich (1969). MR 40-1303.

106 A.J.W. Hilton, A note on edge-colouring multigraphs, Recent Advances in Graph Theory (Ed., M. Fiedler) Academia, Prague (1975) 267-271. MR 52-10473.

107 A.J.W. Hilton, On Vizing's upper bound for the chromatic index of a graph, Cahiers Centre Études Rech. Opér. 17 (1975) 225-233.

108 A.J.W. Hilton, Colouring the edges of a multigraph so that each vertex
         has at most j, or at least j, edges of each colour on
         it, J. London Math. Soc. (2) 12 (1975) 123-128.

109 A.J.W. Hilton, The cover index, the chromatic index and the minimum
         degree of a graph, Proc. British Combinatorial Conf.,
         Aberdeen (Ed., J. Sheehan) Utilitas Math., Winnipeg,
         Man. (1976) 307-317. MR 52-13464.

110 A.J.W. Hilton, Definitions of criticality with respect to edge-colouring,
         J. Graph Theory 1 (1977) 61-68.

111 A.J.W. Hilton, An analogue of Vizing's theorem on the chromatic index
         of a multigraph (to appear).

112 A.J.W. Hilton, A 4-colour conjecture for planar graphs, Proc. Fifth
         Hungarian Conf. on Combinatorics, 1976 (to appear).

113 A.J.W. Hilton, Embedding incomplete latin rectangles and extending the
         edge-colourings of graphs, Nanta Math. (to appear).

114 A.J.W. Hilton, An improved upper bound on the chromatic index of a
         multigraph, Ars Combinatoria (to appear).

115 P.E. Himelwright and J.E. Williamson, On 1-factorability and edge-
         colorability of Cartesian products of graphs, Elem. Math.
         29 (1974) 66-67. MR 50-189.

116 F.C. Holroyd and P.E.D. Strain, Group theory and edge-colourings of
         regular graphs, Pre-print, Open University (1977).

117 L.C. House, A k-critical graph of given density, Amer. Math. Monthly 74
         (1967) 829-831. MR 36-2524.

118 J.L. Hursch Jr., Growth number and colorability of graphs, Notices Amer.
         Math. Soc. (to appear).

119 R. Isaacs, Infinite families of non-trivial trivalent graphs which are
         not Tait colorable, Amer. Math. Monthly 82 (1975) 221-239.
         MR 52-2940.

120 R. Isaacs, Loupekhine's snarks: A bifamily of non-Tait-colorable graphs,
         J. Comb. Theory (B) (to appear).

121 H. Izbicki, Zulässige Kantenfärbungen von pseudo-regulären Graphen 3
          Grades mit der Kantenfarbenzahl 3, Monatsh. Math. $\underline{66}$
          (1962) 424-430. MR 26-2879.

122 H. Izbicki, Zulässige Kantenfärbungen von pseudo-regulären Graphen mit
          minimaler Kantenfarbenzahl, Monatsh. Math. $\underline{67}$ (1963) 25-
          31. MR 26-5557.

123 H. Izbicki, An edge-coloring problem, Proc. Smolenice Symposium on
          Graph Theory (Ed., M. Fiedler) Publ. House Czech. Acad.
          Sci., Prague (1964), Academic Press, New York (1964) 52-
          62. MR 30-656.

124 H. Jaakson, On solutions of a topological two-colour problem, Uč. Zap.
          Tartu Gos. Univ. $\underline{46}$ (1957) 43-62. MR 19-671.

125 H. Jaakson, On solutions of the topological four-colour problem, Tartu
          Riikl. Ül. Toimetised $\underline{102}$ (1961) 263-274. MR 26-4347.

126 F. Jaeger, Sur l'indice chromatique du graphe représentatif des arêtes
          d'un graphe régulier, C. R. Acad. Sci. Paris (A) $\underline{277}$
          (1973) 237-239 = Discrete Math. $\underline{9}$ (1974) 161-172. MR 48-
          146.

127 I.T. Jakobsen, Some remarks on the chromatic index of a graph, Arch.
          Math. (Basel) $\underline{24}$ (1973) 440-448. MR 48-10874.

128 I.T. Jakobsen, On critical graphs with chromatic index 4, Discrete
          Math. $\underline{9}$ (1974) 265-276. MR 50-161.

129 I.T. Jakobsen, On graphs critical with respect to edge-colouring,
          Infinite and Finite Sets, Keszthely, 1973, North-Holland
          (1975) 927-934. MR 52-10474.

130 E.I. Johnson, A proof of four-coloring the edges of a cubic graph, Oper.
          Res. Centre, Univ. California (1963) = Amer. Math. Monthly
          $\underline{73}$ (1966) 52-55. MR 33-934.

131 E. Kaiser and H. Walther, Eine Bemerkung zu einer Arbeit von M.E.
          Watkins über verallgemeinerte Petersensche Graphen, J.
          Comb. Theory (B) $\underline{11}$ (1971) 227-233. MR 44-2662.

132 Ya. Kalnin'sh, The number of ways of colouring the branches of a tree

and a multitree, Cybernetics $\underline{1}$ (1965) 114-116.

133 A.B. Kempe, On the geographical problem of four colours, Amer. J. Math. $\underline{2}$ (1879) 193-200.

134 S. Kitamura, On the edge-chromatic number of a graph, Rep. Fac. Sci. Engrg. Saga Univ. $\underline{3}$ (1975) 17-24. MR 51-2972.

135 D. König, Über Graphen und ihre Anwendung auf Determinantentheorie und Mengenlehre, Math. Ann. $\underline{77}$ (1916) 453-465.

136 D. König, Theorie der Endlichen und Unendlichen Graphen, Akademische Verlagsgesellschaft, Leipzig (1936), Reprinted: Chelsea Publ. Co., New York (1950). MR 12-195.

137 A. Kotzig, Z teorie konečnych pravidelných grafov tretieho a štvrtého stupňa (On the theory of finite regular graphs of degree three and four), Čas. Pěst. Mat. $\underline{82}$ (1957) 76-92. MR 19-876.

138 A. Kotzig, Transformations of edge-colourings of cubic graphs, Discrete Math. $\underline{11}$ (1975) 391-399. MR 52-2942.

139 A. Kotzig, Change graphs of edge-colourings of planar cubic graphs, J. Comb. Theory (B) $\underline{22}$ (1977) 26-30.

140 H. Kronk, M. Radlowski and B. Franen, On the line chromatic number of triangle-free graphs, (Abstract in Graph Theory Newsletter $\underline{3}$ No. 3 (1974) 3).

141 R. Laskar and W. Hare, Chromatic numbers of certain graphs, J. London Math. Soc. (2) $\underline{4}$ (1971) 489-492. MR 45-6680.

142 L. Lesniak, A. Polimeni and D. VanderJagt, Mixed Ramsey numbers: Edge chromatic numbers vs Graphs, Theory and Applications of Graphs, Springer Lecture Notes (to appear).

143 C.R. Marathe, On the dual of a trivalent map, Amer. Math. Monthly $\underline{68}$ (1961) 448-455. MR 23-A3558.

144 M. Mather, The Rugby footballers of Croam, J. Comb. Theory (B) $\underline{20}$ (1976) 62-63. MR 52-10494.

145 C.J.H. McDiarmid, The solution of a time-tabling problem, J. Inst.

144

Maths. Applics. $\underline{9}$ (1972) 23-34. MR 45-9668.

146 P. Medgyessy, Sur la structure des réseaux finis, cubiques et coloriés,
Mathésis $\underline{59}$ (1950) 173-176. MR 12-195.

147 L.S. Mel'nikov, The chromatic class and the location of a graph on a
closed surface, Mat. Zametki $\underline{7}$ (1970) 671-681 / Math.
Notes $\underline{7}$ (1970) 405-411.

148 L.S. Mel'nikov, Some topological classifications of graphs, Recent
Advances in Graph Theory (Ed., M. Fiedler) Academia,
Prague (1975) 365-383.

149 G.H.J. Meredith, Some families of non-Hamiltonian graphs, Combinatorics
(Eds., D.J.A. Welsh and D.R. Woodall) Inst. Math. Appl.,
Southend-on-Sea (1972) 221-228. MR 51-2998.

150 G.H.J. Meredith, Regular n-valent, n-connected, non-Hamiltonian, non-n-
edge-colourable graphs. J. Comb. Theory (B) $\underline{14}$ (1973)
55-60. MR 47-65.

151 G.H.J. Meredith and E.K. Lloyd, The Hamiltonian graphs 0_4 to 0_7,
Combinatorics (Eds., D.J.A. Welsh and D.R. Woodall) Inst.
Math. Appl., Southend-on-Sea (1972) 229-236.

152 G.H.J. Meredith and E.K. Lloyd, The footballers of Croam, J. Comb.
Theory (B) $\underline{15}$ (1973) 161-166. MR 48-149.

153 G.J. Minty, A theorem on three-coloring the edges of a trivalent graph,
J. Comb. Theory $\underline{2}$ (1967) 164-167. MR 34-1031.

154 N. Nakanishi, Quantum field theory and the coloring problem of graphs,
Commun. Math. Phys. $\underline{32}$ (1973) 167-181. MR 48-10327.

155 B.H. Neumann, An embedding theorem for algebraic systems, Proc. London
Math. Soc. (3) $\underline{4}$ (1954) 138-153. MR 17-448.

156 E.A. Nordhaus and J.W. Gaddum, On complementary graphs, Amer. Math.
Monthly $\underline{63}$ (1956) 175-177. MR 17- 1231.

157 O. Ore, The Four Color Problem, Academic Press, New York (1967). MR 36-
74.

158 E. Parker, Edge coloring numbers of some regular graphs, Proc. Amer.

Math. Soc. 37 (1973) 423-424. MR 47-1658.

159 R. Penrose, Applications of negative dimensional tensors, Combinatorial
Mathematics and its Applications (Ed., D.J.A. Welsh)
Academic Press, London (1971) 221-244. MR 43-7372.

160 G. Ringel, Färbungsproblem auf Flächen und Graphen, Mathematische
Monographien 2. VEB Deutscher Verlag der Wissenschaften,
Berlin (1959). MR 22-235.

161 M. Rosenfeld, On Tait coloring of cubic graphs, Combinatorial Structures
and their Applications (Eds., R.K. Guy *et al.*) Gordon and
Breach, New York (1970) 373-376. MR 41-8297.

162 H. Sachs, Regular graphs with given girth and restricted circuits, J.
London Math. Soc. 38 (1963) 423-429. MR 28-1613.

163 D.E. Scheim, The number of edge 3-colorings of a planar cubic graph as
a permanent, Discrete Math. 8 (1974) 377-382. MR 50-163.

164 P. Sentis, Quelques résultats relatifs au coloriage des cartes, C. R.
Acad. Sci. Paris 230 (1950) 355-357. MR 11-377.

165 P.D. Seymour, On multi-colorings of cubic graphs and conjectures of
Fulkerson and Tutte (to appear).

166 C.E. Shannon, A theorem on coloring the lines of a network, J. Math.
Phys. 28 (1949) 148-151. MR 10-728.

167 S. Stahl, Edge-multicolorings of graphs, Pre-print.

168 S. Stahl, Fractional edge-colorings (to appear).

169 J.L. Synge, Two isomorphs of the four-color problem, Canad. J. Math. 19
(1967) 1084-1091. MR 35-989.

170 G. Szekeres, Oriented Tait graphs, J. Austral. Math. Soc. 16 (1973)
328-331. MR 48-10879.

171 G. Szekeres, Polyhedral decompositions of cubic graphs, Bull. Austral.
Math. Soc. 8 (1973) 367-387. MR 48-3785.

172 G. Szekeres, Polyhedral decomposition of trivalent graphs, Combinatorial
Mathematics (Ed., D.A. Holton) Springer Lecture Notes No.
403 (1974) 125-127.

173 G. Szekeres, Non-colourable trivalent graphs, Combinatorial Mathematics
III (Eds., A.P. Street and W.D. Wallis) Springer Lecture
Notes No. 452 (1975) 227-233. MR 51-7928.

174 G. Szekeres and H. Wilf, An inequality for the chromatic number of a
graph, J. Comb. Theory $\underline{4}$ (1968) 1-3.

175 P.G. Tait, [Remarks on the colouring of maps], Proc. Royal Soc. Edin.
$\underline{10}$ (1880) 501-503, 729.

176 P.G. Tait, Note on a theorem in the geometry of position, Trans. Royal
Soc. Edin. $\underline{29}$ (1898) 270-285.

177 A. Thomason, [On uniquely colourable graphs of maximum valency at least
four] (to appear).

178 G. Trevisan, A proposito di un teorema di Petersen, Rend. Mat. Univ.
Padova $\underline{30}$ (1960) 97-100. MR 22-5961.

179 W.T. Tutte, A ring in graph theory, Proc. Cambridge Phil. Soc. $\underline{43}$ (1947)
26-40. MR 8-284.

180 W.T. Tutte, A class of Abelian groups, Canad. J. Math. $\underline{8}$ (1956) 13-28.
MR 17-708.

181 W.T. Tutte, A non-Hamiltonian graph, Canad. Math. Bull. $\underline{3}$ (1960) 1-5.
MR 22-4646.

182 W.T. Tutte, On the colorings of graphs, Canad. Math. Bull. $\underline{4}$ (1961) 157-
160. MR 25-3859.

183 W.T. Tutte, A geometrical version of the four color problem, Combinatorial Mathematics and its Applications (Eds., R.C. Bose
and T.A. Dowling) Univ. N. Carolina Press, Chapel Hill
(1969) 553-560. MR 41-8298.

184 D.W. VanderJagt, Mixed Ramsey numbers (to appear).

185 L. Vigneron, Sur le problème des quatre couleurs: theorie de la
combinaison, C. R. Acad. Sci. Paris $\underline{223}$ (1946) 705-707.
MR 8-164.

186 L. Vigneron, Remarques sur les réseaux cubiques de classe 3 associés au
problème des 4 couleurs, C. R. Acad. Sci. Paris $\underline{223}$ (1946)

770-772. MR 8-164.

187 L. Vigneron, Sur la parité du nombre des composantes de Taît des
         réseaux cubiques sans isthmes associés au problème des
         quatre couleurs, C. R. Acad. Sci. Paris 249 (1959) 2462-
         2464.

188 L. Vigneron, Sur le nombre des composantes de Taît coupant un contour
         fermé tracé sur un graphe cubique coloré associé au
         problème des quatre couleurs, C. R. Acad. Sci. Paris 253
         (1961) 43-45. MR 23-A3687.

189 L. Vigneron, Dénombremant des colorations que l'on peut tracer sur un
         graphe G planaire, cubique sans isthme, connexe, en
         utilisant au plus trois couleurs d'arêtes, Théorie des
         Graphes (Ed., P. Rosenstiehl) Dunod, Paris (1967) 399-400.

190 L. Vigneron, Le problème de l'introduction d'un nouveau pentagone dans
         une carte plane, Cahiers Centre Études Rech. Opér. 15
         (1973) 367-374. MR 50-1956.

191 V.G. Vizing, Об оценке хроматического класса p-графа (On an estimate of
         the chromatic class of a p-graph), Diskret. Analiz 3
         (1964) 25-30. MR 31-856.

192 V.G. Vizing, Хроматический класс мультиграфа (The chromatic class of a
         multigraph), Proc. Third Siberian Conf. on Mathematics
         and Mechanics, Tomsk (1964).

193 V.G. Vizing, Хроматический класс мультиграфов (The chromatic class of
         multigraphs), Doctoral Thesis, Novosibirsk (1965).

194 V.G. Vizing, Хроматический класс мультиграфа (The chromatic class of a
         multigraph), Kibernetika (Kiev) 3 (1965) 29-39 /
         Cybernetics 3 (1965) 32-41. MR 32-7333.

195 V.G. Vizing, Критические графы с данным хроматическим классом (Critical
         graphs with a given chromatic class), Diskret. Analiz 5
         (1965) 9-17. MR 34-17.

196 V.G. Vizing, Some unsolved problems in graph theory, Uspekhi Mat. Nauk
         23 (1968) 117-134 = Russian Math. Surveys 23 (1968) 125-
         142. MR 39-1354.

197 M.E. Watkins, A theorem on Tait colorings with an application to the
generalized Petersen graphs, Proof Techniques in Graph
Theory (Ed., F. Harary) Academic Press, New York (1969)
171-177.

198 M.E. Watkins, A theorem on Tait colorings with an application to the
generalized Petersen graphs, J. Comb. Theory (B) $\underline{6}$ (1969)
152-164. MR 38-4360.

199 D.J.A. Welsh, Combinatorial problems in matroid theory, Combinatorial
Theory and its Applications (Ed., D.J.A. Welsh) Academic
Press, London (1971) 291-306. MR 43-4701.

200 R.J. Wilson, Introduction to Graph Theory, Oliver and Boyd, Edinburgh,
Academic Press, New York (1972). MR 50-9643.

201 R.J. Wilson, Some conjectures on edge-colourings of graphs, Recent
Advances in Graph Theory (Ed., M. Fiedler) Academia,
Prague (1975) 525-528. MR 52-10477.

202 R.J. Wilson, Edge-colourings of critical graphs, Colloquio Internazionale
sulle Teorie Combinatorie (Ed., B. Segre) Accademia
Nazionale Lincei, Rome (1976) 309-311.

203 R.J. Wilson, Edge-colorings of graphs - A survey, Theory and Applications
of Graphs, Springer Lecture Notes (to appear).

204 R.J. Wilson and L.W. Beineke, Three conjectures on critical graphs, Amer.
Math. Monthly $\underline{83}$ (1976) 128-129.

205 B. Zelinka, Edge-colourings of permutation graphs, Mat. Čas. $\underline{23}$ (1973)
193-198. MR 48-10877.

206 A.A. Zykov, Теория Конечных Графов (Theory of Finite Graphs), Nauka,
Novosibirsk (1969). MR 29-986.

Index of names

Index of definitions